IEE TELECOMMUNICATIONS SERIES 39

Series Editors: Professor C. J. Hughes
Professor J. D. Parsons
Professor G. White

QUALITY *of* SERVICE *in* TELECOMMUNICATIONS

Other volumes in this series:

QUALITY *of* SERVICE
in TELECOMMUNICATIONS

A. P. Oodan, K. E. Ward and A. W. Mullee

The Institution of Electrical Engineers

Published by: The Institution of Electrical Engineers, London,
United Kingdom

British Library Cataloguing in Publication Data

A CIP catalogue record for this book
is available from the British Library

ISBN 0 85296 919 8

Typeset by FSH Print & Production Ltd., London
Printed in England by Bookcraft, Bath

Contents

Preface

This book has been written to fill an important gap in the literature on telecommunications. Quality of service is becoming a topic of increasing importance and has reached the stage where it needs to be treated as a subject in its own right. With the intensifying pace of liberalisation of the telecommunications market, coupled with privatisation of incumbent telecommunications operators and the increasingly sophisticated services required by knowledgeable users, quality of service is becoming a significant competitive differentiator. The professional management of quality of service is therefore becoming an important exercise on the part of network and service providers.

This book attempts to clarify the considerable confusion that appears to exist on the understanding of quality of telecommunications services. One of the reasons for this confusion arises from the lack of differentiation between network performance and quality of service as required, delivered and perceived by customers. Yet another reason is the lack of an internationally accepted architectural framework. Many studies have been carried out, some within particular frameworks, others relating to individual aspects of quality and standing alone. Hence, the state of the art does not conform to a logical and coherent approach to the many issues that need to be resolved to obtain a comprehensive understanding of the subject.

Written from a customer perspective, the book aims to identify and discuss the key issues related to quality of service and presents a logical framework for its management. The management of quality brings together many diverse topics such as telecommunications engineering, marketing, statistical techniques, customer care, economics, operations management and total quality management. An in-depth treatment has not been attempted, nor have solutions been offered to many of the problems; the objective has been to provide a sufficient level of basic information to enable the readers to develop solutions to meet their particular needs. It is hoped that the proposed architectural framework will be adopted and refined by an international standards forum as the basis for future studies on the subject.

The target audience for this book includes service and network providers, users and user groups, regulators, equipment manufacturers and students of telecommunications. Telecommunications is characterised by change and this will impact on quality of service requirements. The upcoming information age and the pace of technology development is giving rise to new requirements and means of delivering multimedia information that will erode the traditional telecommunications infrastructure and its value chain. The convergence of the telecommunications, IT, media and entertainment industries is creating an information market with a plethora of new collaborative and competing players. In reviewing some of the studies into future quality of service requirements in order to speculate on what might be required, it is recognised that the only certainty is uncertainty, never-the-less it is believed that the proposed framework and principles outlined in this book are enduring and will form a useful basis for future studies.

Antony Oodan
Keith Ward
Tony Mullee

Foreword

The liberalisation of telecommunication networks and services in the market place of the European Union arrives on 1 January 1998. The removal of exclusive and special rights means that any new supplier can compete with the traditional incumbent operators under fair and nondiscriminatory conditions. The enactment of positive EU regulation, harmonising the diverse national legislation on some key aspects, such as licensing systems, interconnection, open network provision. universal service schemes, voice telephony etc., has facilitated the thrust for the creation of a European internal market. This concerns the full deployment of alternative infrastructure, mobile, satellite, cable TV systems etc. to enhance the development of advanced technologies, as well as deepening the competition for market access in the local loop.

An assessment of this upcoming scenario justifies the publication of this book about the quality of telecommunications and information services. The authors have outlined a complete methodology to define the quality parameters and their perception from the diverse positions of users and suppliers. A liberalised market leads to the differentiation amongst the players competing for the provision of networks and services. The customer satisfaction might be based on minimum quality levels, which would be a matter for regulation; on the other hand, competition may lead to additional quality features offered on market access, intelligent services, efficiency etc. at a convenient price.

The spread of the Information Society is giving more emphasis to new services, sometimes carried over broadband networks or by means of more general connectivity, such as Internet services. In such an environment, defining quality parameters requires a more complex analysis compared with the more basic telecommunications services like voice telephony. This book is particularly useful to explore appropriate methods to this end, and to draw out preliminary conclusions regarding the quality components of 'non-commodity' information services, such as digital multimedia services, delivered over many different physical infrastructures, fixed or mobile.

The book sets out an enlightened linkage between quality, standards, charging, system security etc. Network access and interoperability represent major requirements from the users' satisfaction. The systematic approach in this book has for several months been the basis for a study by a Round Table, within the framework of the Information and Communications Technologies [ICT] Partnership, which has more than sixty user associations, under the European Commission. This study is organised by FITCE (Federation of Telecommunications Engineers of the European Community) and is chaired by one of the authors of this book, Antony Oodan. We are pleased to have the opportunity to participate in this Round Table and appreciate the opportunity to write a foreword to introduce this very relevant contribution to the state of the art on quality of telecommunications services.

Eugenio TRIANA
Directorate-General XIII
European Commission
Brussels
25 July 1997

Acknowledgments

The authors record here the help given by many in the preparation of the manuscript of the book. In particular we wish to thank the following:

Don Cochrane of Phillips Communications for his contribution on QOSMIC project under RACE in Section 3.3, Peter Gibbon of Open-IT Ltd. for commenting on the review of ISO 13236 (Chapter 3), Graham Cook of BT for commenting on the contents of Chapters 3, 10, 11 and 13, Jonathan Pitts and John Shormans of University College, London, for their contribution on ATM performance (Chapter 10), Phil Davidson and Dave Mustil for comments on ITU-T and ETSI and Dave Smith and Alan Penny of BABT for comments on CPE quality matters (Chapter 13), Ernst Weiss of INTUG for contribution on business users' requirements (Chapter 14), Claire Milne for her contributions on quality requirements of residential users and special interest groups (Chapter 14), Sandy Berg of University of Florida for comments on weighted index used in Florida (Chapter 15), Vivian Watkind of National Regulatory Research Institute, Ohio, for discussions on NRRI report in Section 15.4, Peter Scott of European Commission for the contribution in Section 16.4.2, Keith Dickerson for his contribution in Section 18.7 and Peter Haigh of Quality Plus for his contribution on TQM in Appendix 1.

We wish to acknowledge the intellectual copyrights from the following organisations for material used in this book:

Multimedia communications Forum for the material in Section 3.4, the NRRI for material used in summarising the findings in Section 15.4, Florida Public Service Commission, for the performance measures used in Florida in Section 16.3.4, the Federal Communications Commission for performance measures indicated in Section 16.3.4 the OECD for the contents of Table 16.4, the European foundation for Quality Management (EFQM) for the use of their model in Appendix 1 (Section A.1.3), DG XIII of the European commission for permission to use the findings of Round Table 3 in Appendix 4 and ETSI for the information used from ETR 003 in many parts of the book.

Acknowledgments are made to Prof. John Flood for his kind assistance in editing this book.

The Authors

Glossary

AAL	ATM adaption layer
AAR	automatic alternative rerouteing
ABR	answer bid ratio
ACD	agreed commitment date
ACTS	Advanced Communications Technologies and Services
AMPS	analogue mobile phone system
ANSI	American National Standards Institute
AT&T	American Telephone and Telegraph Co.
ATIS	Alliance for Telecommunications Industry Solutions
ATM	asynchronous transfer mode
AUSTEL	Australian Telecommunications Authority
B-ISDN	broadband ISDN
BABT	British Approvals Board for Telecommunications
BCH	bids per circuit per hour
BS	British Standard
BSI	British Standards Institution
BT	British Telecommunications plc
CCITT	Comite Consultatif International des Telegraphes et des Telephones
CDV	cell-delay variation
CEL	cell-error ratio
CENELEC	European Committee for Electrotechnical Standardisation
CIPFA	Chartered Institute of Public Finance and Accountancy
CLASS	customer local-area signalling services
CLI	calling-line identity
CLP	cell-loss probability
CLR	cell-loss ratio
CMR	cell-misinsertion rate
COU	central operations unit
CPE	customer-premises equipment
CRI	Centre for Study of Regulated Industries
CSCW	computer-supported co-operative working

CTD	cell-transfer delay
CTR	Common Technical Regulation
CWC	Cable & Wireless Communications
DAVIC	Digital Audio Visual Council
DBS	direct broadcast satellite
DCE	data-communications equipment
DIS	draft international standard
DP	distribution point
DSD	delay-sensitive data
DTE	data-terminal equipment
DTI	Department of Trade & Industry (UK)
EACEM	European Association of Consumer Electronic Manufacturers
EC	European Commission
EDI	electronic data interchange
EEC	European Economic Community (now known as European Union)
EFQM	European Foundation for Quality Management
EII	European Information Infrastructure
EN	Euro Norm (European Standard)
ETNO	European telecommunications-network operators
ETR	ETSI Telecommunications Report
ETS	ETSI Telecommunications Standard
ETSI	European Telecommunications Standards Institute
EU	European Union
EURESCOM	European Institute for Research and Strategic Studies in Telecommunications
FCC	Federal Communications Commission (USA)
FDM	frequency-division multiplex
FEXT	far-end crosstalk
FIT	failure in item hours
FPLMTS	future public land-mobile telecommunication systems
GATT	General Agreements on Tariffs and Trade
GII	global information infrastructure
GNS	Group of Negotiations on Services
GOS	grade of service
GSM	Group Spécial Mobile
HLSG	High Level Strategy Group
IBC	integrated broadband communications

ICSTIS	Independent Committee for the Supervision of Standards of Telephone Information Services
ICT	information and communications technologies
IDD	international direct dialling
IDN	integrated digital network
IEC	International Electrotechnical Commission
IETF	Internet Engineering Task Force
IFIP	International Federation for Information Processing
IN	intelligent network
INTUG	International Telecommunications User Group
IP	intelligent peripherals
ISDN	integrated-services digital network
ISO	International Standards Organisation
ISP	internet service provider
IT	information technology
ITU	International Telecommunications Union
ITU-T	ITU – Telecommunications Standardising Sector
JV	joint ventures
LE	local exchange
MCI	Microwave Communications Inc.
MDC	multimedia desktop collaboration
MDT	mean down time
ME	main exchange
MFN	most-favoured nation
MISA	Management of Integrated SDH and ATM Networks
MMCF	Multimedia Communications Forum Inc.
MNC	multinational companies
MOSAIC	Methods for Optimisation and Subjective Assessment in Image Communications
MSA	Metropolitan Statistical Area
MTBF	mean time between failures
MTTF	mean time to failure
MTTR	mean time to repair/replace/restore
NC	network congestion
NCC	National Computing Centre
NEXT	near-end crosstalk
NFU	network field unit
NHSTUG	National Health Service Telecommunications User Group
NICC	Network Interfaces Co-ordination Committee

NII	national information infrastructure
NNI	network-node interface
NOP	National Opinion Poll
NOU	network-operations unit
NP	network performance
NRRI	National Regulatory Research Institute
NSTS	network-services test system
NTP	network terminating point
NTT	Nippon Telegraph and Telephone Corporation
ODP	open distributed processing
OECD	Organisation for Economic Co-operation and Development
OFTEL	Office of Telecommunications (UK)
OLO	other licensed operator
OLR	overall loudness rating
ONA	open network architecture
ONP	open network provision
OPTUS	a service provider in Australia
OSI	Open Systems Interconnection
PBX	private branch exchange
PC	private circuits or personal computers
PCB	public call boxes
PCM	pulse-code modulation
PCP	primary crossconnect points
PMR	private mobile radio
POTS	plain old telephony service
PSDS	packet switched
PSTN	public switched telephony network
PTO	public telephone operator
QDU	quantisation distortion unit
QoS	quality of service
QOSMIC	quality of service verification methodology and tools for integrated broadband communications
QUOVADIS	quality of video and audio for digital television services
RACE	Research and technology development in Advanced Communications technologies in Europe
RBOC	regional Bell operating company
RLR	received loudness rating
RPI	retail price index
SCP	service control point

SDH	synchronised digital hierarchy
SDL	specification and description language
SECBR	severely errored cell block ratio
SERVQUAL	branded abbreviation for SERvice QUALity
SLA	service-level agreement
SLR	send loudness rating
SP	service provider
SPECIAL	Service Provisioning Environment for Consumers' Interactive Applications
SPI	Strategic Planning Institute
SS	signalling system
SSP	service switching point
STMR	side-tone masking rating
STP	signalling transfer point
TAPESTRIES	The Application of Psychological Evaluation to Systems and Technologies in Remote Imaging and Entertainment Services
TCCA	time-critical communications architecture
TELCO	telecommunications company
TELSTRA	the largest service provider in Australia
TMA	Telecommunications Managers' Association
TMN	telecommunications management network
TO	telecommunications operator
TOMQAT	Total Management of Service Quality for Multimedia Applications on IBC Infrastructure
TQM	total quality management
TUA	Telecommunications Users' Association
ULE	user lost erlang
UN	
UNI	user-network interface
VANS	value-added networks
VCR	video cam recorder
VFM	value for money
VOD	video on demand
VPN	virtual private network
VSO	voluntary standards organisations
WIPO	World Intellectual Property Organisation
WTO	World Trade Organisation

Chapter 1
Introduction to quality of service

1.1 Definition of quality

Before attempting to understand the issues related to quality of service in telecommunications, it will be helpful to focus on what is understood by quality and the meaning implied by this term in the context of quality of service (QoS). Many definitions have been used in economics literature, but still there seems to be no single clear definition. Many of these definitions have been reviewed in a report published in Australia [1]. A review of some of the theories is included within this section. For example, Dhrymes [2] defines quality as 'the set of identifiable characteristics exhibited by a given product'. The British Standards Institution (BSI) [3] defines quality as 'the totality of features and characteristics of a product or service that bear on its ability to satisfy stated or implied needs'. Others such as Dorfman and Steiner [4] define quality as 'any aspect of a product...which influences the demand curve'. Leffler [5] states that 'quality refers to the amounts of the un-priced attributes contained within each unit of a priced attribute'.

Garvin [6] identifies five major approaches to the definition of quality in the academic literature from economics, marketing, operations management and philosophy. They are classified as the transcendent, product-based, user-based, manufacturing based and value-based approaches. Garvin concludes that each approach is unique and imprecise in describing the basic elements of product quality. He therefore proposes a framework for thinking about the basic elements of quality in terms of eight dimensions:

(i)   performance (primary product characteristics);
(ii)   features (secondary product characteristics);
(iii) reliability (probability of product failing);
(iv)   conformance (degree to which design and operating characteristics match pre-established standards);
(v)   durability (product life);
(vi)   serviceability (speed, courtesy and competence of repair);
(vii) aesthetics (how the product looks, feels sounds, tastes or smells);
(viii) perceived quality.

Garvin states that each of the five approaches to the definition of quality implicitly focuses on two or three of these dimensions The approaches in the economics literature are identified as product-based (focused on performance, features and durability) and user-based (focused on aesthetics and perceived quality). Each definition from the economics literature is to some extent affected by the focus of the study for which it was developed. However, the definitions generally incorporate the concept of 'characteristics' and consider quality in the context of a particular product.

Lancaster's [7–11] theory of consumer demand emphasised the characteristics or attributes of products. Under Lancaster's approach, a product is viewed as a bundle of characteristics. Consumers derive utility from the characteristics embodied in the product rather than from the product *per se*. Characteristics are defined as 'those objective properties of things that are relevant to choice by people'. Subsequent work by various authors indicates that, for analytical purposes, individual characteristics may be disaggregated into subcharacteristics and several characteristics may be aggregated into an aspect of a service quality. For example, several characteristics of airline services, such as cabin temperature and humidity, aircraft stability, level of crowding and seat comfort may be aggregated into the aspect of on-board comfort. Similarly, the characteristic of seat comfort may be disaggregated into subcharacteristics such as cushion softness, seat width, seat pitch, reclining features and fabric smoothness.

Each product potentially possesses a large number of characteristics but the operational use of Lancaster's model depends on the ability to confine the analysis to a relatively small number of characteristics with measurable properties. Lancaster therefore proposes that practical studies should be limited to the relevant characteristics. Identification of the characteristics and assigning values would then lead to the definition of quality. The characteristics may be expressed quantitatively or qualitatively.

All these definitions in established literature refer to quality in terms of characteristics or attributes associated with a product or service, i.e. things which can be measured. However, Crosby [12] defines quality simply as 'conformance to requirements', this shows that there may be more to quality than can easily be measured.

The role of customer's[†] perception is a major consideration in the specification of quality. It is difficult to understand how customers perceive the product offering and, more important, how they evaluate the quality of the product offering. Some authors when writing on quality suggest that customers evaluate products in a relatively coarse fashion with overall judgements being perhaps more properly described in terms of their attitude to the product. Additionally, it is also important to make the distinction between quality and satisfaction.

Satisfaction is associated with the entire customer product-ownership experience. This is extremely important because satisfied or highly satisfied customers will become loyal customers, and loyal customers will allow the

† Customer is a bill-paying user

companies which serve them to outgrow and outearn their less proficient competitors. Satisfaction will only occur when customers' negative perceptions associated with a product are disconfirmed and thus, in general, this variable tends to be confined to particular transactions, such as using the product for the first time. However, quality as viewed from the customers' perspective, can only be achieved through repeated incidents of satisfaction.

Quality as perceived by customers stems from a comparison of what they feel the product should offer (i.e. drawn from their expectations) with their perceptions of the actual performance of the product. When customers buy a product they will have an expectation of how that product will perform and this will cover a whole host of criteria including:

- conformance to specification;
- performance (primary product characteristics);
- reliability (probability of the product failing);
- availability (probability of the product being in stock);
- simplicity (ease of use of the product);
- durability (product life);
- serviceability (speed, courtesy and competence of repair); and
- aesthetics (how the product looks, feels, tastes, smells and sounds) [6].

Expectations may be considered as predictions, sometimes subconscious, which customers make about what is likely to happen during the consumption of a particular product. In practice, this may differ significantly from the view of the manufacturer or service provider, whose view of quality will be based on their perspective of the product offering. However, the most important evaluation of quality is that which is carried out by the customer. This evaluation draws on perceptions and expectations (covered in Chapter 8) and reinforces Gronroos [13] in defining quality within the service environment as:

$$\text{quality} = \text{customers' expectations} - \text{customers' perceptions}$$

This is reinforced and can be further subdivided using the disconfirmation model as shown in Figure 1.1.

Empirical evidence suggests that there are a number of key characteristics or attributes which customers will generally evaluate to determine the quality of any particular product or service. The disconfirmation model shows that customers' satisfaction will be dependent on both the size and direction of disconfirmation, with only three possible outcomes. When 'perceived' is greater than 'expected', customers will be satisfied; when 'perceived' is equal to 'expected', customers will be neutral, neither satisfied nor dissatisfied (i.e. the product or service is performing exactly as expected); when 'perceived' is less than 'expected', customers will be dissatisfied. The scale of customer satisfaction or dissatisfaction will be dependent on the difference between 'expected' and 'perceived' performance of these key characteristics or attributes.

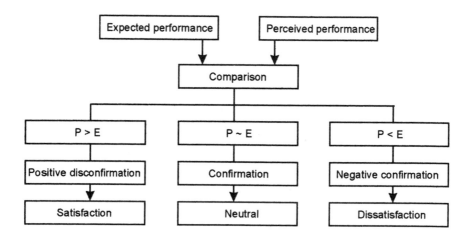

Figure 1.1   Disconfirmation model of customer satisfaction [14]

If quality is 'conformance to requirements' [12] and the real judge of quality is the customer, then they must evaluate quality in relation to their satisfaction (Figure 1.1). Satisfaction is related to repeated-use or entire-ownership experience. This suggests that quality must therefore be related to the product/service life cycle, as shown in Figure 1.2.

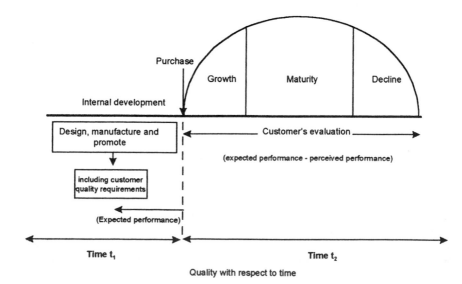

Figure 1.2   Quality related to product life cycle

Product or service quality can change over time. Initially during time t_1 the quality of the product or service will be driven by the development specification, which for a new product or service will be based around customer assumptions of quality, with or without the practical experience of product or service use. During time t_2 this may change with product or service familiarity and usage.

Quality affects customers in two distinct ways. First, quality will affect their initial purchasing decision. Factors affecting purchase will be company image, similar product offerings and product expectation raised by advertising and publicity (Chapter 8). Secondly, during the life of the product or service, quality will be influenced more by the technical performance of the product or service and the customer-support service experienced with that product or service (service surround, discussed in Chapter 9). Customer satisfaction will be positive, negative or neutral, based on the product or service in-life performance. Therefore, quality must be considered with respect to the product/service life cycle.

Considering the arguments above, the definition of quality of service used in this book is as follows:

Quality of service (QoS) is defined in the context of the customer/user or the provider. In the context of the customer or the user it is defined by the attributes or criteria which are considered to be essential in the use of the service. In the context of the service provider, QoS is defined by parameters which contribute towards the end-to-end performance of the service, this end-to-end performance reflecting customers' requirements.

The above definition leans heavily on the work of Lancaster (the concept of attributes that are of benefit to the customer) and Crosby (conformance to customer requirements throughout the entire product/service life cycle).

The principal characteristics of QoS of a telecommunications service are:

(a) parameters which express quality provide benefit to the user and the service provider;
(b) quality should be expressed on a service-by-service basis;
(c) quality parameters could be different depending on the viewpoint taken (i.e. customer or service provider); and
(d) for the customer (and to an extent the service provider) quality parameters are expressed on an end-to-end basis.

Note that, in the past, network performance has been mistaken for QoS. Quality is experienced by the customer and customers state what levels of performance are required. They also state what levels of performance are perceived for services used. Network performance is the technical performance of elements of the network or of the whole network. Network performance contributes towards QoS. Network performance should be derived from QoS to be offered to customers and incorporated into planning schedules. To offer customers the QoS based on the capabilities and limitations of the network is to offer them what is available irrespective of what is required.

1.2 Why quality is important

Although it is hard to pin down the determinants of quality, its importance to industry and consumers is unequivocal. Research has demonstrated the strategic benefits of quality in contributing to market share and return on investment as well as in lowering manufacturing costs and giving improvements in productivity [15]. Thus quality has become a prime issue in the operations and strategies of many organisations. Therefore, the dilemma facing most managers today is 'do I get competitive advantage through quality and if so how much advantage do I get and for how long?' Is the growing interest in quality a fad or the reshaping of corporate goals? Evidence suggests that, by focusing on quality, companies can increase market share. This is evident with the rise to fame of the Japanese companies such as Honda, JVC, Hitachi and Toyota which were once renowned for shoddy second-class goods but after embracing quality turned around the entire perception of their product. The Japanese, by concentrating on quality, increased their market share by providing reliable goods at reduced prices. It has also been proved that a quality approach based on customer needs will allow a company to gain and maintain a price premium over its competition. This in turn provides the basis for increased market share with the consequential positive impact on revenue and profitability. This view is illustrated by the work of Lee and Hsiang who developed a model which describes the impact of product quality on sales [16]. Product quality is postulated to affect the customers repeat buying and also 'word-of-mouth' product promotion.

1.3 Quality and customer behaviour

Quality covers more than just price or features, it covers the perception of that product throughout its entire life, i.e. its reliability. Extra features of a product which meets customers' needs can often influence sales more than price alone. This was typified by the Japanese car industry which was the first to add more features such as 'radios' to the basic car specification and achieved a higher market share even though its basic prices were higher. However, additional features alone are not the only answer; the car still needs to be reliable and perform throughout its entire working life to a set of base requirements as perceived by the customer. Therefore, quality and reliability are two different but closely related aspects of the provision of better products to customers. A customer buys a product on the probability that it will function satisfactorily throughout its entire working life, provided that it is maintained and used correctly. Reliability is therefore the proof of quality; if a product is used and performs correctly throughout its entire working life, then by definition it is a high quality product.

However, the central tenet of this argument concerns the provision of a 'level' of quality which is closely matched to customers' requirements. This does not imply a situation where a uniform quality is offered to all customers, market sectors or segments of a market. This in turn gives rise to the issues concerning the attributes of the product on which individual customers base their purchasing decisions. All customers, even in a small market segment, are individuals; and what makes each one buy may be very different. Customers may buy based on past experience, recommendation, published literature including the product specification, service-provider reputation, brand and price. Studies have shown that customers do not evaluate quality in isolation and that service providers need to understand the strong coupling between product quality and price [17]. In practice, this leads to the concept of 'value'. Value can be described as a conceptual relationship rather than a concise mathematical equality and represented as:

value = function {benefits (quality) and price}

In Figure 1.3 along a fixed value contour, assuming a competitive environment, it is possible to identify discrete product quality categories. The first, the 'economy offering', is of lower quality than similar products and is therefore offered at a lower price. This is usually aimed at the lower end of

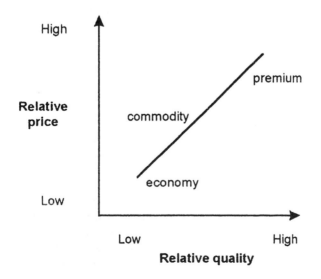

Figure 1.3   The value contour

the market and the service provider will have settled for a low profit margin and rely on volume sales to remain profitable. The second is the 'commodity offering' which will be of similar price and quality to other competitive products, the service provider relying upon other attributes to sell the product, such as company or brand name. Alternatively, the market may consist of only a few dominant service providers (such as in the oil industry and communication industry) where the service providers will attempt to differentiate their product in some way. The final category, here referred to as the 'premium offering' stands out as having a higher quality than similar products and is consequently offered at a premium (higher) price.

The service provider will then exploit the product quality to enhance sales; however, the premium quality will only appeal to a smaller number of customers with specific requirements for higher quality and with the perception that they are still achieving 'value'.

1.4 Quality of service in telecommunications

The telecommunications industry is undergoing tremendous change. The environment can be characterised by liberalisation, deregulation, global expansion, new technological advances and new advanced products and services. Global liberalisation and deregulation have opened up new market opportunities for expansion but with the corresponding threat of competition in once-monopolistic secure markets. The old traditional approach, focused on technology advancement and delivery to a secure market which was firmly under the control of the telecommunications-service provider, is no longer valid. The industry was driven by technology and not customer requirements, which led to the controlled growth in basic tele-communications services. Today, especially in the industrialised countries, these services show deceleration in growth and even signs of saturation which has produced a fundamental change in the telecommunications industry, i.e. the search for new value-added services and new growth markets. Companies in this new industry environment will have to operate closer to the market and customer needs. The new successful breed of telecommunications company will need to adapt very quickly to changes in the market, using technology innovatively to meet customer needs.

In the future, innovation will be important when attempting to do business in new and existing markets. The global telecommunications market is a market with a wide spectrum of needs ranging from basic communication requirements in the developing regions to very complex requirements in the developed world. A telecommunications company which wishes to compete in this market usually has the option to 'cherry pick', i.e. service the profitable customers. The service provider can offer them customised levels of quality. Therefore, the main telecommunication drivers of the future will come from the customers, customers groups, user groups, standards bodies and regulators. The existing and emergent roles of the standards bodies, user groups and regulators are the subject of Chapters 13, 14 and 15 of this book.

In the past, the quality of telecommunications services has been determined by the quality level offered by the monopoly service providers. Within the UK market and increasingly in other markets, competition is becoming more widespread and for the first time customers will have a choice in their service provider. In the future, customer needs are likely to be the main driver in the telecommunications industry. This is likely to become especially prevalent as the global economy becomes increasingly service orientated with competition well established. When competition exists, customers can set aside old loyalties and choose the service provider which best serves their needs. Potential and actual competitors will have a vital interest in identifying areas of quality where there is potential to gain or sustain competitive advantage and subsequently set their quality levels and prices accordingly. In this competitive environment, it will be essential to ensure that each customer is getting value for money. The concept of 'value' has already been identified as being a major factor in determining customers' purchasing preferences, and a major component of value is quality.

Customers are also beginning to realise that they have a major role to play in setting quality levels within the telecommunication industry and also in determining what new services should be introduced. Customers are beginning to come together to form 'user groups' or 'organisations', thus increasing the volume and power of their voice within the industry. Many such organisations are in existence today and their role in setting customer requirements is not at present fully appreciated by service providers. User groups or groups of customers will, by virtue of their individual and combined purchasing power, play an ever increasing role in the determination of telecommunications requirements.

1.5 Measuring quality

When measuring quality there are two areas which need to be addressed: how to measure quality and what aspects of quality are to be measured.

The definition of quality developed for use in this book (in Section 1.1) implies that conformance to customer requirements throughout the entire product/service life cycle should be considered in the management of quality. Hence the measurements ought to reflect the various stages of the service, from its inception, through launch and its life until it is withdrawn. It is, however, difficult to define measures of quality which can be uniformly agreed across all interested parties, e.g. customers, service providers, regulators, user groups and standards bodies.

There is also some debate within the telecommunications industry as to what precisely constitutes quality. From the customer's perspective it can only be the end-to-end quality plus the surrounding service quality (i.e. service surround). By this is meant the totality of the quality criteria. To achieve this, a holistic approach to quality must be adopted.

Despite these difficulties, there are two distinct areas of quality which customers evaluate, covering the purchase decision and the in-life performance. Both these can be split into two main measurement areas, one 'technical' in terms of how a product or service performs (i.e. what a customer really receives) and the other 'functional', how the product or service is provided and performs throughout its entire life cycle; this also includes all interaction between the customer and service provider as well as brand image [18]. Therefore, when looking at how and what to measure, it is important to take an holistic approach and consider the entire life cycle.

It is helpful to think of quality in terms of quality indicators. A quality indicator can be either subjective (measuring customers' perceptions) or objective (measuring the amounts of characteristics relevant to individual aspects of service quality). To obtain a true measure of quality, it is necessary to obtain both sets of indicators. However, owing to the customised nature of quality, it is necessary to obtain key indicators which are of most importance to customers, and base all measurements around these key indicators.

1.6 Conclusions

Quality affects everyone and their perception of quality is personal, thus making its effective management complex. Quality is related to value but covers more than just price and features, relating rather to the entire ownership experience. The quality criteria of services in telecommunications are expressed in quantitative and qualitative terms. Lessons can be learnt from other industries in the relationship between customer behaviour and quality. Quality studies in the telecommunications industry should be based on market requirements and mapped to the service providers' capability. However, it must also be understood that the perception of quality may differ between a customer and service provider and therefore, quality must be broken down between these different viewpoints for effective management. Before management of quality is dealt with, a review of the key issues within the telecommunications industry is carried out in Chapter 2.

1.7 References

1 'Quality of service: conceptual issues and telecommunications case study'. Bureau of Transport and Communications Economics, Australian Government Publishing Service, Canberra, 1992
2 DHRYMES, P. J.: 'Price and quality changes in consumer capital goods: an empirical study' *in* GRILICHES, Z. (Ed.): 'Price indexes and quality change: studies in new methods of measurement' (Harvard University Press, Cambridge 1971)
3 BS EN ISO 8402: 'Quality management and quality assurance — vocabulary' (British Standards Institution, 1995)

4 DORFMAN, R. and STEINER, P. O.: 'Optimal advertising and optimal quality', *Am. Rev.*, 1954, pp. 826–836

5 LEFFLER, K. B.: 'Ambiguous changes in product quality', *Am. Rev.*, 1982, **72**, (5), pp. 956–967

6 GARVIN, D. A.: 'What does "product quality" really mean?', *Sloan Manage. Rev.*, Fall 1984, pp. 25–43

7 LANCASTER, K. J.: 'A new approach to consumer theory', *J. Political Economy*, April 1996, pp. 132–157

8 LANCASTER, K. J.: 'Consumer demand: a new approach' (Columbia University Press, New York, 1971)

9 LANCASTER, K. J.: 'The measurement of changes in quality', *Rev. Income Wealth*, June 1977, pp. 157–172

10 LANCASTER, K. J.: 'Variety, equity, and efficiency' (Columbia University Press, New York, 1979)

11 LANCASTER, K. J.: 'The economics of product variety: A survey', *Marketing Sci.*, 1980, **9**, (3), pp. 189–206

12 CROSBY, P. B.: 'Quality without tears, the art of hassle-free management' (McGraw Hill, 1984)

13 GRONROOS, C.: 'Strategic management and marketing in the service sector' (Helsingfors Swedish School of Economics and Business Administration, 1982)

14 WALKER, J. L.: 'Service encounter satisfaction: conceptualised', *J. Services Marketing*, 1995, **9**, (1), pp. 5–14

15 PARASURAMAN, A., ZEITHAML, V. A., and BERRY, L. L.: 'A conceptual model of service quality and its implications for future research', *J. Marketing*, Fall 1985, **49**, pp. 41–50

16 LEE. H. L., and HSIANG, T. C.: 'A cost model with sales effects'. Proceedings of *Western Regional Conference, American Institute for Decision Sciences*, Monterey, 1985, pp. 122–124

17 DEVEREUX, G. D.: 'Using quality as a strategic weapon'. *IEEE Globecom*, 1985, paper 45.6

18 MORGAN, N. A., and PIERCY, N. F.: *Indust. Marketing Manage.*, 1992, **21**, pp. 111–118

Chapter 2
Key issues of quality of service

2.1 Introduction

An overview of the key issues associated with the study of QoS in telecommunications should assist the reader to appreciate the scope of topics to be covered in this book. The identification of most, if not all, of the key issues may be facilitated by a matrix made up of the parties involved in the supply, use and monitoring of these services separated into 'national' and 'international'. The quality issues may be different in character for 'national' and 'international' services; hence telecommunication services are conveniently divided into these categories. The matrix is shown in Table 2.1.

Table 2.1   Key-issues matrix

Parties	Services	
	National	International
Network provider	Section 2.2	Section 2.2
Service provider	Section 2.3	Section 2.3
Equipment manufacturer	Section 2.4	Section 2.4
Customer/user	Section 2.5	Section 2.5
Standardising bodies	Section 2.6	Section 2.6
Regulator	Section 2.7	Section 2.7

The principal parties in the supply, use and monitoring of tele-communication services are the network provider, service provider, equipment manufacturer, user, standards bodies and regulator. Other bodies such as universities and research establishments may be considered as playing supporting roles to any one (or more) of the above players.

2.2 Network provider

The issues related to QoS that should be addressed by the network provider on a national basis are:

(a) aspects related to service-level agreements (SLAs) between customers and service providers. The issues related to SLAs are discussed in Chapter 6;

(b) where interconnection exists, aspects related to SLAs with other network providers; and

(c) the specification of quality-related technical performance of network systems required of equipment suppliers.

In the international area, the network provider has to address issues on quality in the following areas:

(d) relevant standards on performance necessary in interconnection with foreign networks; and

(e) SLAs with foreign network providers which may or may not be additional to the requirements of the standards.

2.3 Service provider

In the provision of national telecommunication services, the service provider has to address issues related to QoS in the following areas:

(a) management of QoS indicators for:

- internal business use,
- the benefit of customers,
- meeting regulator's requirements, and
- differentiating in a competitive market;

(b) general management of quality, including total quality management (TQM) and how management of QoS fits in to the overall TQM (see Appendix 1 for further information on TQM);

(c) influence on the standardising bodies for optimum benefit to service provision;

(d) support of research into quality issues;

(e) determination of customers' requirements on QoS and its management;

(f) determination of customers' perception of QoS and its management;

(g) specific needs on quality of special-interest groups, e.g. the disabled, low-income groups etc.;

(h) economics of QoS; and

(i) societal aspects.

For international services the following specific issues have to be addressed:

(j) QoS matters in SLAs with foreign administrations on the correspondent provision of service between host and foreign country;

(k) QoS aspects in relation to competitors in the global market, where a global service provider is bypassing the traditional bilateral-correspondent relationships for the provision of international services;

(*l*) SLAs with multinational companies; and

(*m*) influence of standards bodies on quality-related matters for international services.

(a) QoS indicators

For network providers which are also service providers, the network-related performance parameters are usually measured by the network providers within the network systems. Network providers which are not service providers have to supply the service provider with relevant network-related performance for the service provided. For non-network-related performance parameters, the service providers have had to institute their own performance systems. The principal shortcoming in this area has been the choice of performance parameters which are mainly of interest to providers rather than customers. Choice of QoS indicators to meet providers' needs and additional indicators, where necessary, to meet customers' requirements should satisfy all parties.

(b) Total quality management

Management of QoS would be most effective if carried out within the context of a company-wide TQM process and not in isolation. There is adequate information available on the topic of TQM to enable any company to apply the fundamental principles and to benefit from its implementation. TQM is discussed briefly in Appendix 1.

(c) Service provider and standards bodies

The service provider has to address the issue of when to involve standardising bodies to develop performance standards for new services. It also has to address the issue of what types of performance parameter are to be standardised. A current shortcoming is a tendency to focus on the technical issues without taking full account of their influence on customers' requirements.

(d) Research on quality issues

With the development of new applications leading to increasingly sophisticated services, it became necessary for more performance standards to be specified. With increased complexity in technology, the parameters became more complicated. Many of the service providers in the OECD (Organisation for Economic Co-operation and Development) countries had their own research departments which carried out fundamental research into many aspects of telecommunications, including quality. Nowadays, following privatisation, the research establishments are increasingly being run on a commercial basis. They are funded to address specific issues to improve the commercial viability of the service providers by better exploitation of technology. Nevertheless, studies on quality are being carried out, sometimes for the mutual benefit of many providers. An example is the European Institute for Research and Strategic Studies in Telecommunications (EURESCOM). This organisation has the common interests of participating European network and service operators. However, the research establishments have also concentrated on the service providers' interests, despite some excellent work for the benefit of customers. There has been criticism from users that insufficient has been done to meet their interests.

(e) Determination of customers' QoS requirements

Determination of customers' QoS requirements is a necessary activity on the part of the service provider, especially if it wants to provide the quality that will satisfy customer needs. Following the emergence of competition outside the USA, there is an increased awareness of the customers' QoS requirements. Despite this awareness, there is no formal mechanism to capture these, nor an agreed methodology to manage this exercise.

(f) Customer perceptions of quality

Surveys of customer perceptions of quality have been carried out for a long time. However, there is a criticism, with some justification, that the perceptions have been measured to enhance the image of the service provider. It would be helpful to develop a set of indicators meaningful to customers, in addition to those required by the service provider to monitor the internal performance of the organisation.

(g) Quality issues of special interest groups

In many of the more developed countries, the needs of special-interest groups, e.g. the disabled, low-income groups etc., have to be identified and addressed, especially by the dominant service provider, often at the request of the regulator or to enhance the image of the service provider.

(h) Economics of quality

The cost of providing a level of quality, the revenue generated and the effects of competition are some of the considerations that should be addressed by service providers. No qualitative measures have been developed to determine the balance of QoS benefit and cost.

(i) Societal aspects

The dominant service provider in any country is most likely to face the societal issues. These include provision of unprofitable services to remote areas, attitudes towards the countryside, arts and heritage of the country, sponsorship of sporting events, commitment towards the community, education of the younger generation etc.

The issues under *(j)*, *(k)*, *(l)* and *(m)* for international services are similar in nature to the corresponding ones in the national scene, but with different implications.

2.4 Equipment manufacturers

The following are principal areas where equipment manufacturers should be involved with QoS issues:

(a) design and manufacture of individual elements of telecommunications networks in order to meet specified end-to-end performance;

(b) design of test equipment and support systems to monitor performance of individual elements of a telecommunication system and, in certain cases, of delivered end-to-end performance;

(c) influencing standardising bodies on realistic performance standards.

The quality issues to be addressed by the manufacturers are unlikely to be significantly different for national and international provision of services. Their principal roles are to meet the performance statements specified by the purchasers, the service providers and the network providers. Additionally, they could enlighten the standards bodies on what is realistically achievable and to influence the timing of standards for the common benefit of all parties in the telecommunications industry.

2.5 Customers/users

In relation to national telecommunication services, users are likely to be concerned with the following issues on quality:

(a) publication, by service providers, of delivered QoS on parameters which are meaningful to customers;

(b) quality needs of specific groups being met by the service provider. Examples of specific groups are:

- banks and certain newspaper publishers which require zero outage during certain periods of the day;
- broadcasters which require large call-handling capacity from service providers during periods of televoting; and
- quality requirements of the disabled, low-income groups etc.

(c) businesses which require customised performance in contractually binding SLAs with service providers and network providers;

(d) consultation by network and service providers on their requirements of performance.

In the provision of services in the international scene the following quality issues will be of concern to users:

(e) SLAs between multinationals and global alliances of service providers on quality aspects. These will be more complex than SLAs for services provided within a country; and

(f) consultations with users and user groups on their quality requirements for sophisticated services.

The principal shortcomings in the above areas are the lack of adequate representation of the customers' interests by the service providers.

2.6 Standardising bodies

There are national, regional and international standardising bodies. Although there is only one international standardising body for telecommunications, the Telecommunications Standardising Sector of the International Telecommunications Union (ITU-T), the convergence of telecommunications, information technology (IT) and information industries is resulting in other international standards organisations influencing the QoS delivered over telecommunications networks, e.g. the International Standards Organisation (ISO). The issues to be addressed by the standards bodies are:

(a) development of end-to-end performance standards for an international connection and determining how the national part can be deduced from this; and

(b) development of standards for specific issues, e.g. relationships between degrading factors on transmission quality and their effects on customers' perceptions.

2.7 Regulators

Where regulators exist, their influence is confined to within the boundary of a country. The only current exception is the European Commission, but the World Trade Organisation (WTO) is beginning to take an interest in global telecommunications regulation. The key issues on QoS to be addressed by the national regulators are:

(a) the influences on quality under monopoly and competitive environments;

(b) determination of QoS parameters on which delivered quality data are to be published by the service and network providers;

(c) audit of published QoS data;

(d)   whether there ought to be a 'reward and punishment' policy towards providers in relation to delivered quality;

(e)   the timing of introduction of quality parameters for new services; and

(f)   steps to maintain delivered quality by service providers.

The regulators appear not to concern themselves with international services. However, the European Commission does have an intra-Community interest in the harmonisation within the European Union member countries. This concerns international services between service providers in different member countries of the European Union.

Chapter 3
Review of major studies

3.1 Introduction

Before undertaking the treatment of the QoS in telecommunications, it is useful to review briefly some of the principal studies related to QoS in telecommunications. A large number of studies have been carried out in various parts of the world on various aspects of QoS. Studies are carried out within the research departments of the service providers, organisations representing regulators, other organisations specifically assigned to the study of specific issues in telecommunication and many other bodies. The International Federation for Information Processing (IFIP), which has a strong representation from universities around the world, had its fourth workshop on QoS in March 1996 [1] and has reported many studies on QoS. Even though much useful work has been undertaken, there is a lack of an architectural framework for the study of QoS in telecommunications. More importantly, there is a degree of confusion between QoS and network performance.

It will be impossible to carry out a thorough review of all studies. However, it is pertinent to summarise some major developments on QoS. Three studies have been reviewed, namely:

- a project called QOSMIC under the RACE collaborative research programme sponsored by the European Commission;
- the work of the Multimedia Communication Forum;
- the draft material scheduled to become ISO standard 13236 on 'Framework for QoS in Information Technology'.

3.2 European collaborative research in telecommunications

Some research into QoS aspects of advanced services and networks is being carried out within the European collaborative telecommunications research projects sponsored by the European Commission. The Research and technology development in Advanced Communications technologies in Europe (RACE) programmes contained, for example:

(a) LUSI (R2092) 'Likeable and Usable Service Interfaces: This project addressed the needs of the general public for user–service interfaces so that services will not be rejected because they are unusable. Many user trials have been carried out on interfaces for a variety of retrieval, messaging and conversational services. The output from the project has been a set of guidelines for user–service interfaces and the results have been included in standards formulated by the European Telecommunications Standards Institute (ETSI).

(b) MOSAIC (R2111) 'Methods for Optimisation and Subjective Assessment in Image Communications': The main objective was to develop and prove reliable methods for assessing picture quality and degradation. This will enable developers to evaluate their image-transmission systems and compare, in a meaningful manner, the relative performance of various systems.

(c) NEMESYS (R1005) Traffic and QoS Management for Integrated Broadband Communications (IBC): The project majored on the telecommunications management network† (TMN) requirements for ATM (asynchronous transfer mode) traffic and QoS management. It developed and carried out trials of a number of testbeds for ATM network simulation, together with service and network-management applications.

(d) QOSMIC (R1082) 'Quality of Service Verification Methodology and Tools for Integrated Broadband Communications': See section 3.3.

(e) TOMQAT (R2116), 'Total Management of Service Quality for Multimedia Applications on IBC Infrastructure': The project identified requirements for the management of QoS and also the ATM-performance parameters to be measured. An exhaustive list of QoS parameters and QoS management functions was developed. Methods have been specified for performance-parameter measurement and QoS assessment for the service provider and user, taking into account a multiservice environment.

The RACE programme has now been completed and the fourth framework programme, Advanced Communications Technologies and Services (ACTS), is underway and includes a number of projects where QoS implications are dealt with, for example:

(a) QUOVADIS (AC056) 'Quality Of Video and Audio for Digital television Services': The project concerns the management of QoS in digital television operation covered by the *MPEG 2* standard and involving a multiprovider, multioperator and multinational environment. It aims to identify QoS parameters and examine means for automatically controlling and supervising the performance of the overall digital television distribution over the chain from service providers to end users. It is intended to develop and carry out trials of QoS-management prototypes.

† TMN (Telecommunication Management Network) is the ITU-T name for the group of computer support systems used to manage the network and its services. The related standards are contained in the ITU-T *Recommendation M.3200* series.

(b) SPECIAL (AC0912), 'Service Provisioning Environment for Consumers' Interactive Applications': Although primarily concerned with concepts of accounting and tariffing-based billing for a variety of multimedia services in a broadband environment, it also aims to develop and validate customer-care concepts. It is intended to test the results in a pilot project with a substantial number of end users.

(c) TAPESTRIES (ACO55) 'The Application of Psychological Evaluation to Systems and Technologies in Remote Imaging and Entertainment Services': The objective is to develop an understanding of the key psychological factors which contribute to customer acceptance of new entertainment and information services. Based on the analysis of psychological criteria for specific service applications of new entertainment media and display technologies, it is intended to produce a framework for determining acceptability criteria.

(d) TELE-SHOPPE (P032) Teleshopping services using virtual reality and interactive multimedia: The main objectives are 'to investigate how the use of advanced multimedia technologies and virtual reality can simulate the "touch and feel" of physical shopping in a telepresence shopping experience and to incorporate this research in a series of demonstrators and field trials of tele-shopping services using broadband ISDN (integrated services digital network) and ATM networks with the aim of measuring usability of the user-interface design for such network services'. These user-interface designs will be tested in usability trials.

(e) MISA (AC080) 'Management of Integrated SDH and ATM Networks': The objective of this project is 'to realise and validate via field trials the optimum integrated end-to-end management of hybrid SDH (synchronous digital hierachy) and ATM networks in the framework of the open-network-provision (ONP) environment'. It aims to deal with a global broadband-connectivity-management (GBCM) service where network operators provide an end-to-end connectivity service to service providers and business customers in a multiprovider and co-operative-service-management environment. An important requirement of the study is to produce proposals on the management of the QoS agreed between the end user and service provider, which relies on the performance of those networks which are used to provide the end-to-end connectivity.

(f) Sub-area 5: Service Access for Advanced Multimedia Applications: Designing open, flexible and user-friendly access systems for new multimedia services. The introduction of interactive multimedia services requires intelligent interfaces capable of assisting the user to understand the services available and access them, locate the required material, contact other parties and retrieve, store and manipulate information. This has been recognised as an area where significant advances are required, and the following ACTS projects deal with aspects of the problem: AC018 SMASH, AC032 SETBIS, P-104 DVBIRD, AC010 MUSIST, AC030 KIMSAC, AC033 SOMMIT, AC071 SICMA and AC082 DIANE.

The next collaborative research programme, Framework 5, was formulated during 1997 by the European Commission with the co-operation of the European Union member states.

3.3 QOSMIC study

QOSMIC was a RACE I project (R1082) and ran from 1990 to 1992. It produced both theoretical and practical work on QoS and provided a valuable contribution to the subject. The work was originally aimed to benefit integrated broadband communications (IBC). A more detailed description of the QOSMIC work is contained in a report [2]. The theoretical work included development of the 'timeline' and 'brick-wall' models described below.

3.3.1 Timeline model

It is difficult to reconcile such diverse QoS parameters as information error rate, help-desk behaviour and accuracy of billing. Clearly, they are all QoS parameters, but it was considered that there were fundamental differences between them. As more QoS parameters were examined, these difficulties became worse. To solve this problem, the QOSMIC 'timeline' model was developed.

The timeline model is illustrated in Figure 3.1. This model shows that there is a hierarchy of timescales involved in the interactions between a customer (user) and a service provider. There are three timelines in the model: per call, per customer and per service infrastructure. This model has the benefit

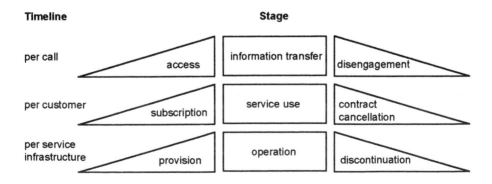

Timeline | Stage

per call — access — information transfer — disengagement

per customer — subscription — service use — contract cancellation

per service infrastructure — provision — operation — discontinuation

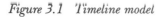

Figure 3.1 Timeline model

of putting such concepts as QoS of billing and customer services alongside such QoS parameters as bit-error rate, and of simplifying these dissimilar concepts.

(a) Per-call timeline

The top horizontal timeline of Figure 3.1 represents a single call. Here the user of the service is involved. Using the concepts of ITU Recommendation E.810 [3], it is meaningful to examine the stages of a single use of a service in chronological phases. Thus, a single call has an 'access phase', an 'information transfer phase' and a 'disengagement phase'. These are characteristic of a connection-oriented architecture, which is the assumed basic structure of the IBC service, i.e. possessing all of the three call stages (as opposed to a connectionless mode of operation).

- The access phase covers the user indicating that use of the service is required, identifying to the provider the destination of the call and negotiating any parameters of the call (bandwidth, reverse charging, closed user group etc.). The provider either makes the connection or fails to do so (due to incorrect number, network congestion, equipment failure etc.) and the recipient indicates call acceptance or rejection.
- Information transfer covers the period from the destination user accepting the call until either party indicates that disengagement is required.
- Disengagement concludes the process, generating and storing a charging record for that call. In the ISDN, a user's terminal could have more than one call in progress simultaneously.

Each of these phases has its own QoS and service provider's parameters.

(b) Per-customer contract timeline

A single call occupies only a small portion of the time period during which a customer subscribes to a service (of the order of seconds or minutes), but there will be many such calls and possibly some in parallel. This relationship is shown by the middle horizontal timeline in Figure 3.1. Note that this timeline applies to both the customer and the user, who are not necessarily the same person. This timeline covers the provision of a service to a single customer, it has three phases in a similar manner to the per-call timeline. These are:

- 'subscription', e.g. connecting the user to the network, and on to specific service provided by the service provider;
- 'service use', which is the normal in-service state where a service is available to the user on demand; and

- 'contract cancellation' when a user moves premises or the contract between provider and customer is otherwise ended.

This timescale spans days, months or years.
   Examples of QoS for the middle timeline would be parameters such as:

- time to provide service;
- time to repair totally-failed service;
- accuracy of directory-service entry for customer;
- accuracy for provider acknowledging bill payment (e.g. claiming that a bill was unpaid and discontinuing service, when the bill had actually been paid on time;
- training of users; and
- help-desk availability.

(c) Service-infrastructure timeline

The lowest horizontal timeline of Figure 3.1 has a still longer timescale than the customer timeline and may be up to tens of years. This is the entire lifetime of a service, from 'provision' (initial planning and installation of the necessary equipment to provide the service), through 'operation' (within which there will be many sequential and parallel instances of the middle timeline) and finally 'discontinuation' when the service is replaced or obsolete (e.g. the UK Inland Telegram service).

3.3.2 Brick-wall model

Given that IBC is a wide-area multiservice environment with multiple providers of basic services (bearer services) as well as a numerous value-added-service providers, services delivered to users may be assembled from different service components from different suppliers. Thus there is interworking between services, both on a higher-layer-to-lower-layer basis, e.g. an e-mail service supported by an X.25 data service, and on a peer-to-peer basis, e.g. interconnected national networks. In functionality terms, these different services (i.e. service components of the service delivered to the user) will both receive and provide QoS on the bases of peer and higher-layer-to-lower-layer relationships.
   In the brick-wall model, each of these service components is represented by a single 'brick', as shown in Figure 3.2. The complete service is represented by an assembly of these bricks. The brick-wall model assists the understanding of the relationships between the services, both between layers and peer-to-peer.
   The brick-wall model:

- permits the examination of each single service 'brick' as a component in the provision of the final QoS delivered to the end user;

- relates the performance of the service, as measured by the provider, to received QoS and the functionality of the service; and
- facilitates examination of the composite effect of the QoS of multiple services on the QoS to the end user, whatever the relationship between the services which make up the final service.

This service 'brick' concept can be applied to both the OSI model[†] (e.g. transport, session etc.) and ITU-T communications concepts such as the ATM adaptation layer (AAL) [4]. Therefore, it is possible to reconcile the OSI and ITU views of communications, i.e. functional layers separated by service-access points against physical units separated by reference points. In addition, it can also be used to deal with the service provided by a physical object such as a switching node.

The single brick illustrates the propagation of QoS through a service layer. It has two possible classes of QoS input and two of QoS output, as shown in Figure 3.2. All inputs and outputs may not always be present, because some services do not operate in an environment with peers. There may also be occurrences of a single QoS parameter from a single service, e.g. a distribution service.

The inputs to the brick are:

- the Peer QoS received from peer providers, i.e. providers at the same OSI level and providing a similar service to the provider in question; and
- the QoS from lower-level providers which make an underlying service available that the provider in question needs in order to function.

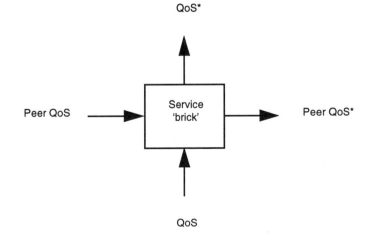

Figure 3.2   The QOSMIC brick model

† Open Systems Interconnection (OSI) is a concept originally developed to allow communications between computers and terminals from different manufacturers. The specifications and protocols were developed by the International Standards Organisation (OSI), based on a seven-layer protocol known as the ISO Reference Model for OSI.

The outputs from the brick are:

- the Peer QoS* given by a provider to its peers, i.e. it is their received peer QoS. Note that this may be the same brick that is providing peer QoS to this brick, if the relationship is mutual (as for both ends of the transport layer); and
- QoS* is the QoS provided to the overlaid services, i.e. it is their received QoS from this lower layer.

Contained within the brick is its functionality which provides its service, together with its performance which maps to the provided QoS* and peer QoS*. The functionality may be changed by either of the received QoSs and this, in turn, may affect its performance and hence one or both of the provided QoS*s.

A service supplied to an end user can require the combined effect of multiple bricks, stacked to form a 'brick wall'. There are conversions from system performance to QoS and *vice versa* as the quality of the underlying brick is modified, changed and extended by subsequent bricks. By means of this model, the effect of all components of a service can be taken into account.

3.4 Studies by the Multimedia Communications Forum

An example of the progress in defining future QoS requirements is the work being carried out by the Multimedia Communications Forum Inc. (MMCF) [5]. The MMCF used the high-level 3 × 3 matrix in the ITU-T E.430 [6] and its studies covered the shaded part illustrated in Figure 3.3.

	Speed	Accuracy	Dependability
Connection set-up (access)			
User information transfer			
Connection Disengagement (release)			

Shaded portion show scope of MMCF's work on QoS

Figure 3.3   Framework from ITU-T E.430

The multimedia applications illustrated in Table 3.1 were largely derived from ITU-T I.211 [7] and the Forum's earlier work.

Table 3.1   Multimedia communications applications [7]

Application category	Example of application
Interactive:	
Multimedia collaboration	Videotelephony/ videoconference Distance learning Remote presentation Audiographics Videosurveillance
Multimedia information retrieval	Video retrieval Image retrieval Document retrieval Data retrieval Vidotex
Multimedia mail	Videomail Voicemail Document mail
Distribution	Existing-quality-TV distribution High-definition-TV distribution Pay TV Document and image distribution Audio-information distribution Digital-information distribution Video-information distribution Full-channel broadcast videography

One of the most onerous applications for which QoS requirements will need to be satisfied is that of 'multimedia-desktop-collaboration' (MDC) teleservices, sometimes known as computer-supported co-operative working' (CSCW). In its report on 'multimedia communications QoS' [5], the MMCF defines the primary† parameters for MDC-user information transfer as:

(i) *Speed*
Audio: audio transfer delay — which can interfere with the dynamics of voice communications and increase the effects of echo.

Video: video/audio differential delay — which is defined as the video minus audio delay and describes the difference in delay between the video and audio media as perceived by the user. The delay may be fixed or variable. The

† ITU-T *Recommendation I.350* defines primary QoS parameters as those determined on the basis of direct observations of events at service access points or connection-portion boundaries. A derived parameter is determined on the basis of observed values of one or more relevant primary parameters.

effect is less noticeable under low or variable frame rates where the video does not convey continuous motion. From a study by Kurita *et al.* [8] it appears that 50% of users detect a differential delay of ± 250 ms and user sensitivity increases to ± 100 ms with high frame rates.

Delay-sensitive data (DSD): DSD/audio differential delay — describes the difference in delay between DSD and audio media, as perceived by the user, where DSD includes pointer, control or echoplex information.

Data: data rate — defined as the data-transfer rate for user-available data and includes text, still pictures, facsimile, binary files and formatted document information.

(ii) *Accuracy*

Audio:

• audio-frequency range — the bandwidths can comply with ITU-T Recommendation G.711 for 3.1 kHz PSTN or higher, depending on the application, i.e. users may demand better voice quality when video is also present;
• audio level — should be similar to those for conventional telephone equipment as laid down in ITU-T Recommendations P.31 and P.34;
• audio error-free interval — errors can be perceived by users as 'clicks' and 'pops' and this interval is the time between such audio impairments.

Video:

• video frame rate — is the rate at which a single frame of video (sometimes composed of two interlaced fields) is presented;
• video resolution — specifies the ability to distinguish video details in the spatial dimension;
• video error-free interval — is the typical time between transmission-caused user-perceived video impairments, e.g. blocking, colour distortion, edge jerkiness etc.

Delay-sensitive data: DSD error-free interval — is the typical time between user-perceived data impairments caused by transmission errors, e.g. erroneous pointer displacements and text errors.

The above parameters would be meaningless to most users' perception of QoS, particularly for video impairments. The American National Standards Institute (ANSI) has defined some of the user-perceived video impairments [9], as shown in Table 3.2. The user's perception for a typical multimedia teleconferencing application with video and accompanying audio, together with text and images, will be influenced by many aspects including the satisfaction of the interactive session, previous experience, ease of use and the environment in which the system is evaluated. For such complex situations it is difficult to predict what user expectations might be or how they might be mapped onto network performance parameters.

Table 3.2 Video-performance terms and definitions [5]

Name	Description
Block distortion	Distortion of the received image, characterised by the appearance of an underlying block-encoding structure
Blurring	A global distortion over the entire image, characterised by reduced sharpness of edges and spatial detail
Edge busyness	Distortion concentrated at the edge of objects, characterised by temporally varying sharpness (shimmering) or spatially varying noise
Error blocks	A form of block distortion where one or more blocks in the received image bear no resemblance to the current or previous scene and often contrast greatly with adjacent blocks
Jerkiness	Motion which was originally smooth and continuous is perceived as a series of distinct 'snapshots'
Mosquito noise	Distortion, typically seen around the edges of moving objects, characterised by moving artifacts around edges and/or blotchy noise patterns superimposed over the objects (e.g. a mosquito flying around a person's head and shoulders)
Scene-cut response	Perceived impairments associated with a scene cut
Smearing	A localised distortion over a subregion of the image, characterised by reduced sharpness of edges and spatial detail

The 1995 report of the MMCF also defines the performance parameters, for MDC teleservices (see Table 3.3), for a number of QoS classes, namely:

QoS class 0:   not defined
QoS class 1:   for basic multimedia applications
QoS class 2:   for enhanced multimedia applications
QoS class 3:   for premium multimedia applications

The MMCF has also defined the performance parameters relating to these classes for bearer services [i.e. between the end user-to-network interfaces, (UNIs)], public and private networks and terminal equipment.

Table 3.3   Quality-of-service classes for MDC teleservices [5]

QoS parameter	QoS class 1	QoS class 2	QoS class 3
Audio-transfer delay with echo control†	<400 ms	<400 ms	<400 ms
Audio-frequency range	≥0.3 – 3.4 kHz	≥0.3 – 3.4 kHz	≥0.05 – 6.8 kHz
Audio level (typical)	– 20 dBm0	– 20 dBm0	– 20 dBm0
Audio error-free interval	>5 min	>15 min	>30 min
Video transfer delay	<10 s (still image only)	<600 ms	<250 ms
Video/audio differential delay	N/A	>–400 and <200 ms	>–150 and <100 ms
Video frame rate	N/A	≥5 frames/s	≥25 frames/s
Video resolution	N/A	≥176 x 144	≥352 x 288
Video error-free interval	N/A	>15 min	>30 min
DSD/audio differential delay	<1 s	<200 ms	<100 ms
DSD error-free interval	>5 min	>15 min	>30 min
Data rate	≥5 kbit/s	≥50 kbit/s	≥500 kbit/s

†   An audio-transfer-delay limit of 25 ms is applicable when supporting connections to conventional telephones without supplementary echo control.
MDC teleservices using satellite facilities may exceed 400 ms; however, the 400 ms limit is still recommended

The MMCF has defined a number of multimedia-desktop-collaboration scenarios with the relevant user expectations and these are shown in the MDC scenario matrix of Table 3.4 where levels 1–3 of the user scenarios correspond to QoS classes 1–3.

Table 3.4   User expectations by media type for MDC scenarios [5]

Levels	Motion video	Audio	Text	Graphics/ animation	Images
Level 1	None	PSTN quality	Noticeable delay Jerky delivery	Single frame: —VGA colour —noticeable delay No animation	Single image: —significant delay
Level 2	Talking heads Jerky delivery Lossy compression	PSTN quality Apparent sync. with video	No perceptible delay Steady delivery	Near real-time delivery of slide-show-presentation graphics motion between slides	Single image: noticeable delay
Level 3	Comparable with VHS tape quality Multiple video channels	Two or more channels Lip sync. with video Comparable with FM broadcast stereo	No perceptible delay Steady delivery	No perceptible delay of presentation graphics Cartoon-like animation	Single image: —near real-time delivery
Level 4	Range of: — broadcast — studio quality — HDTV Multiple video channels	3 or more channels Surround sound comparable with digital-CD audio	No perceptible delay Steady delivery	No perceptible delay of presentation graphics Cartoon-like animation	Single image: —no perceptible delay

3.5 ISO quality of service framework for applications in information technology

The ISO has produced a draft international standard (*ISO/IEC DIS 13236*) 'Information technology — QoS — framework' [10], which was circulated for comment in 1996. The purpose 'is to provide a common basis for the co-ordinated development and enhancement of a wide range of standards that specify or reference quality-of-service (QoS) requirements or mechanisms'. The initial work in developing the framework was done in the context of OSI with the objective of supplementing and clarifying the description of QoS contained in the Basic Reference Model of OSI as described in ITU-T *Recommendation X.200* and *ISO/IEC 7498-1*. The framework is structured in such a way as to make it possible for other communities to adopt its approach, concepts, terminology and definitions. The framework is a structured collection of concepts and their relationships which describes QoS and

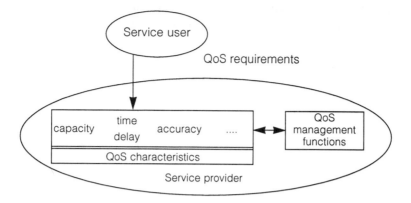

Figure 3.4   Relationships between QoS concepts

enables the partitioning of, and relationships between, the topics relevant to QoS to be expressed by a common description. It is intended to assist developers of new and revised standards which define and use QoS mechanisms; and users expressing requirements for QoS.

The relationships between the fundamental QoS concepts are illustrated in Figure 3.4.

QoS characteristics represent some aspect of the QoS of a system, service or resource which can be identified and quantified. The framework defines some 54 characteristics under groupings of: time-related, coherence, capacity-related, integrity-related, safety-related, security-related, reliability-related and other characteristics. As an example, under time-related the 'lifetime characteristic' has been defined as 'the period of time for which the data are valid' and may be quantified in any time unit.

User requirements drive the QoS-management activities and originate with an application process that wishes to use a service. The requirements may be retained in an entity which may also analyse them to generate further requirements that are conveyed to other entities as QoS parameters, and so on. Examples of such parameters are:

- a desired level of characteristic, e.g. a target;
- a maximum or minimum level of characteristic, e.g. a bound;
- a measured value, used to convey historical information;
- a threshold level;
- a warning to take corrective action; or
- a request for operations on managed objects relating to QoS, or the results of such an operation.

QoS management includes functions to meet the needs of the users and applications. These functions may require many different types of action to be performed: negotiation, admission control and monitoring, for example.

These are called QoS mechanisms. The management functions may be defined in a structure of phases of the QoS activity. Examples of such activities are:

- establishment of QoS for a set of QoS characteristics;
- monitoring of the observed values of QoS;
- maintenance of the actual QoS as close as possible to the target QoS;
- control of QoS targets;
- enquiry on some QoS information or action;
- alerts as a result of some event relating to QoS management.

The activities can again be categorised into three phases: prediction phase, establishment phase and operational phase.

In addition to the above, the model contains 'general QoS mechanisms' dealing with the three phases, QoS monitoring, QoS maintenance and the QoS verification, and conformance, consistency and compliance.

3.6 Conclusions

Considerable activities have taken place and continue to take place on the topic of QoS. The lack of an agreed architectural framework has resulted in a vast number of *ad hoc*, unco-ordinated and 'stand-alone' studies. They do not fit together and the combined outputs of these studies lack synergy. For example, Lancaster University have developed a quality-of-service architecture (QoS-A) which offers a framework to specify and implement required performance properties of multimedia applications over integrated-service networks in a heterogeneous environment [11]. The various studies would have been more beneficial if they fitted to an architectural framework for the understanding of the subject. This fact is now becoming obvious to many parties. An internationally agreed framework for QoS will greatly benefit network providers, service providers and customers. Chapter 4 proposes a framework for the study and management of QoS in a methodical manner and later chapters deal with specific issues related to these. Further chapters deal with QoS issues outside the framework.

3.7 References

1   'Quality of service — description, modelling and management'. Proceedings of the Fourth International IFIP Workshop on Quality of Service, Paris, March, 1996
2   RACE: 'QoS verification methodologics'. RACE project R1082 QOSMIC – deliverable 1.4C, 1992
3   ITU-T E.810: 'Framework of the recommendations on the serveability performance and service integrity for telecommunications service' (International Telecommunications Union, 1992)
4   PITTS, J. M., and SCHORMANS, J. A.: 'Introduction to ATM design and performance: with applications analysis software' (John Wiley and Sons, 1996)

5   MULTIMEDIA COMMUNICATIONS FORUM, INC.: 'Multimedia communi-
    cations quality of service'. MMCF/95-010, June 1995
6   ITU-T E.430: 'Quality of service framework' (International Telecommunications
    Union, 1992)
7   ITU-T I.211: 'B-ISDN service aspects' (International Telecommunications Union,
    1993)
8   KURITA, T., IAI, S., and KATAWAKI, N.: 'Effects of transmission delay in
    audiovisual communication', *Electron. Commun. Jap.*, Part 1, 1994, **77**, (3)
9   'T1A1 proposed draft ANSI standard: specification of video performance terms
    and definitions' (American National Standards Institute) Refer to Reference 5.
10 ISO/IEC DIS 13236: 'Information technology – quality of service – framework'
    (International Standards Organisation, 1996)
11 HUTCHINSON, D., MAUTHE, A. and YEADON, N.: 'Quality of service
    architecture: monitoring and control of multimedia communications', *IEE
    Electron. Commun. J.*, **9**, (3), June 1997, pp.100–106

Chapter 4

A framework for quality-of-service management

4.1 Introduction

The conclusions of Chapter 3 lead to the recognition of the need for a practical framework to facilitate the study of QoS in telecommunications. In this chapter a framework to meet this need is described; it will be discussed in more detail in subsequent chapters.

Any framework, to be useful, must meet the following requirements;

(a) it must identify and describe all principal components (functions or activities);
(b) it must explain the principal characteristics of each component from which one could identify the key issues;
(c) it must provide guidelines for the management of the key issues; and
(d) it must be practical to achieve the objective for a wide spectrum of users.

The framework described in this chapter meets the above requirements. It is developed from ETSI document *ETR 003 Edition 2* on quality of service [1]. Its components are:

(a) a quality cycle;
(b) a matrix to facilitate the determination of QoS criteria; and
(c) a methodology for the study and management of QoS.

Components of the framework are illustrated in Figure 4.1.

The quality cycle, described in more detail in Section 4.2 describes four viewpoints of QoS and their relationship with each other. The matrix to facilitate the determination of QoS criteria forms the principal building block for one of the four viewpoints, namely customer's QoS requirements, and is

Quality cycle	Matrix to determine QoS criteria	Methodology for the study and management of QOS

Figure 4.1   Components of the framework for the study and management of QoS

described in Section 4.3. The matrix can be used for the study of the other three viewpoints whenever necessary. The methodology for the management of the QoS cycle describes how the overall management of QoS fits together, and is dealt with in Section 4.4.

4.2 The quality cycle

4.2.1 Principal parties

There are two principal parties: the customers and the service providers. The QoS studies may be further divided into the future and past. For the service provider, such division leads to planned and achieved quality. For the customers this division leads to their QoS requirements or expectations and their perception of experienced performance.

This concept is illustrated in Figure 4.2.

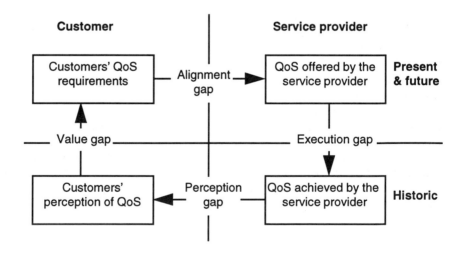

Figure 4.2   Quality cycle

4.2.2 Customers' QoS requirements

QoS requirements by the customer are the statement of parameters and the level of quality of a particular service. The level of quality may be expressed by the customer in technical or nontechnical language. A typical customer is not concerned with how a particular service is provided, or with any of the aspects of the network's internal design, but only with the resulting end-to-end service quality. From the customer's point of view, QoS is expressed by parameters, which·

- focus on user-perceived effects, rather than their causes within the network;
- do not depend in their definition on assumptions about the internal design of the network;
- take into account all aspects of the service from the customer's point of view;
- may be assured to a user by the service providers, sometimes in contractual terms; and
- are described in network-independent terms and create a common language understandable by both the user and the service provider.

For example, a customer may indicate that the acceptable number of occasions when moderate difficulty in call clarity during a telephone conversation is tolerated would be a maximum of one in 100 calls. The service provider translates this requirement to network-performance parameters and assigns target values to network elements to result in an end-to-end performance of not more than 1 call in 100 experiencing moderate difficulty in call clarity.

4.2.3 QoS offered by the service provider

QoS offered by the service provider is a statement of the level of quality expected to be offered to the customer by the service provider. The level of quality is expressed by values assigned to QoS parameters. Each service would have its own set of QoS parameters. The service provider may express the offered QoS in nontechnical terms, where necessary, for the benefit of customers, and in technical terms for use within the business.

For example, a service provider may state, for the benefit of customers, that the availability of basic telephony service is expected to be 99.995% in a year with not more than a 15 min break at any one occasion and not more than three breaks over the year. In this case the technical description is the same.

4.2.4 QoS achieved by the service provider

QoS achieved by the service provider is a statement of the level of quality provided to the customer. This is expressed by values assigned to parameters, which should be the same as specified for the offered QoS so that the two can be compared. These performance figures are summarised for specified periods of time, e.g. for the previous three months. For example, the service provider may state that the achieved availability for the previous quarter was 99.95% with five breaks of service of which one lasted 65 min.

4.2.5 QoS perceived by the customer

QoS perceived by the users or customers is a statement expressing the level of quality which they 'believe' they have experienced. The perceived QoS is usually expressed in terms of degrees of satisfaction and not in technical terms. Perceived QoS is assessed by customer surveys and from customers' own comments on levels of service. For example, a customer may state that on an unacceptable number of occasions there was difficulty in getting through the network to make a call and may give it a rating of 2 on a five-point scale, with 5 indicating excellent service.

The QoS should include all aspects of quality, not merely the network-related aspects. Customers cannot be asked to state their network-related and non-network-related QoS criteria separately. QoS performance figures would be more meaningful if specified on a service-by-service basis. Some parameters would have the same definition when applied to other services, e.g. provision of service. Other parameters, especially technical ones, could vary in definitions and targets and may have to be specified for each service.

4.2.6 Inter-relationships within the quality cycle

The customer's QoS requirements may be considered as the logical starting point. A set of customer's QoS requirements may be treated in isolation as far as its capture is concerned. This requirement is an input to the service provider for the determination of the QoS to be offered or planned. The service provider may not always be in a position to offer customers the level of QoS they require i.e. alignment gap. Considerations such as cost of quality, strategic aspects of the service provider's business, benchmarking (world's best) and other factors will influence the level of quality offered. The customer's requirements may also influence what monitoring systems are to be instituted for the determination of achieved QoS for the purpose of regular reports on achieved quality. The gap between the delivered and offered quality, called the 'execution gap', is of interest to the service provider. Customers' perception of achieved quality is expressed in terms of satisfaction indices, usually obtained by surveys. The gap between perceived and achieved quality, i.e. perception gap, is of interest to the service provider. In addition, there is a value gap between the customers' original requirements and the perceived QoS.

The combination of inter-relationships and the quality cycle forms the basis of a practical and effective management of quality of telecommunications services.

4.3 Matrix to facilitate determination of QoS criteria

4.3.1 Background and description

A telecommunication service may be subdivided into uniquely identifiable 'service functions' whose sum constitutes the service. The choice and grouping of these service functions are influenced by the sequence in the subscription to a service, i.e. starting from sales and ending with the cessation of the service. The technical, charging/billing and network/service-management functions are grouped separately. The service functions are shown in Table 4.1.

Table 4.1   Service functions making up a telecommunication service

Service management	1 Sales, precontract activities 2 Provision 3 Alteration 4 Service support 5 Repair 6 Cessation
Connection quality	7 Connection establishment 8 Information transfer 9 Connection release
	10 Charging and billing 11 Network/service management by customer

Quality criteria of concern to customers for telecommunication services may similarly be found under the following categories:

- speed
- accuracy
- availability
- reliability
- security
- simplicity
- flexibility

When a matrix is formed with the above sets of criteria, the resulting cells indicate quality-of-service criteria for a telecommunication service. This matrix is illustrated in Figure 4.3. The basic principles of the matrix were first published by Dvorak and Richters [2]. It is perhaps worth noting that a modified version of this work has been used by the National Regulatory Research Institute (NRRI) in its research to identify the key issues on QoS in the USA [3].

These quality criteria have been found to cover most, if not all, quality aspects of a telecommunication service. The authors believe the service functions which have been developed to be most appropriate for representing the various components of a telecommunications service. Since the quality criteria are believed to be those that are most likely to concern customers and the service is subdivided into uniquely identifiable service functions, it should, in theory, be possible to identify all possible QoS criteria most likely to concern customers. This is the basis for identification of all QoS criteria for a telecommunication service. The tendency in the past has been for telecommunication-network engineers to treat network quality criteria in isolation and separate them from other criteria. However, from the customer's point of view, this separation is not desirable.

It is unreasonable to expect customers to separate the network- and non-network-related quality criteria and to state the respective quality requirements separately. Confusion can also arise if separate customer surveys require customers to state the relative importance of technical and nontechnical quality-of-service criteria.

4.3.2 Service functions in the matrix

Service management

The first six service functions introduced in Table 4.1 are grouped under the title 'service management' in the matrix and may be described as follows:

(i) *Sales and pre-contract activities:* In the business relationship between a service provider and the customer for supply, provision and maintenance of a service the first stage is the 'sale'. At this stage, the customer is provided with information on the service features, its limitations, any planned changes in the service features or capabilities for the future, the initial cost to supply, rental (if applicable), maintenance arrangements and the level of quality that can be expected.

All information transactions regarding a service, before the customer agrees to hire a service from a service provider should be included under the service function of 'sales and precontract activities'.

(ii) *Provision:* Provision is the setting up, by the service provider, of all the components required for the operation and maintenance of a service, as defined in the service specification and for use by the customers.

(iii) *Alteration:* During the life cycle of a business relationship between the customer and the service provider, there could be occasions when the customer requires amendments to the service or service features. Such changes or alterations are more frequently demanded by business customers than by residential customers. These changes are best incorporated into the original contract for the service. All activities, such as contract amendment and logistics for the changes to service, are grouped together under this service function.

(iv) *Service support:* The following are included in service support:

- enquiries on the operation of a service (e.g. queries on configuring a personal computer for access to the Internet service);
- documentation on service operation and any other relevant matter,
- procedures for making and progressing complaints; and
- preventive maintenance.

(v) *Repair:* This service function includes all activities associated with repair from the instant a service does not offer one or more of the specified features to the instant these features are restored for use by the customer.

(vi) *Cessation:* In the 'product life cycle' of a business relationship between the customer and a service provider, the last logical service function is the

Service function / Service quality criteria	Speed 1	Accuracy 2	Availability 3	Reliability 4	Security 5	Simplicity 6	Flexibility 7
Service management — sales & precontract activities 1							
provision 2							
alteration 3							
service support 4							
repair 5							
cessation 6							
Connection quality — connection establishment 7							
information transfer 8							
connection release 9							
Charging & billing 10							
Network/service management by customer 11							

Figure 4.3   Matrix to facilitate determination of QoS criteria

cessation of the service. All contract clauses come to an end with the cessation of a service. Activities associated with this are: removal of the equipment from customer's premises, settling of accounts and updating of appropriate records.

Connection quality functions

The next three service functions are grouped under the title 'connection quality' and cover the technical quality of the connection. They are:

(vii) *Connection establishment:* All activities associated with the establishment of a telecommunication service are included in this service function. It is the time elapsed from the instant the customer requests service (e.g. lifting of the handset of telephone) to the instant any of the following indications are received:

- ring tone or equivalent;
- destination-engaged tone;
- call-answered indication;
- any other signal indicating the status of the called party or the network.

(viii) *Information transfer:* This service function, probably the most sensitive to customers, groups together all the quality measures that influenced the transfer of information, covering the period from the instant the connection is established to the instant release is requested by the customer.

(ix) *Connection release:* All activities associated with the release of the connection from the instant release is requested by the customer to the instant the connection is restored to its dormant state are covered in this service function. Performance measures include conformance to period specified (within limits) for release and the order in which the release is carried out by the different elements in the network.

(x) *Charging and billing:* This service function deals with all aspects associated with charging and billing. Examples of its constituent parts are: accuracy of charging, retrieval of charging information, format of bills and specific requirements of customers on bill structure, and payments requirements (e.g. by direct debit etc.).

(xi) *Network/service management by the customer:* With increasing sophistication in telecommunication services some control of the service is given to the customer, e.g. the control of routing arrangement for an advanced freephone service where a single, nongeographic (0800) number is used for multiple destinations with caller choice controlled by additional digits keyed under voice prompting. All performance measures associated with the service management by the customer are included in this service function. It also includes all activities associated with the customer's control of predefined changes to telecommunication services or network configurations.

4.3.3 Service-quality criteria in the matrix

The service-quality criteria introduced in Section 4.3.1 may be described as follows:

(a) Speed: This quality criterion is the speed with which one or more of the functions of the service is carried out.

(b) Accuracy: This quality criterion deals with the fidelity and completeness in carrying out the communication function with respect to a reference level.

(c) Availability: Availability is the likelihood with which the relevant components of the service function can be accessed for the benefit of the customer at the instant of request (not be confused with the ITU-T definitions on 'Availability').

(d) Reliability: Reliability is the probability that the service function will perform within the specified limits for speed, accuracy or availability for a given period, e.g. one year.

(e) Security: This quality criterion deals with the confidentiality with which the service function is carried out. No information is to be supplied to an unintended party, nor can information be changed by an unintended party.

(f) Simplicity: The ease of application of the service function. This quality criterion concerns the ergonomic aspects of how the service feature is dealt by the service provider. It also includes the customer's preferred requirements for a particular service. For example customers may prefer a standard, simple and logical 'log-on' procedure for access to an electronic-mail service.

(g) Flexibility: This quality criterion groups together the customer's optional requirements associated with the service. Examples of this are choice of monthly or quarterly telephone bills, the time of the month when monthly bills are sent to the customer, the amount and format of billing information to the business customer etc.

4.3.4 Application of the matrix

Generic cell descriptions for the 77 cells are given in Appendix 2. The principal application of the matrix and the resulting cell descriptions is the derivation of relevant service-specific QoS criteria or parameters for any telecommunications service. An example of the derivation of QoS criteria and parameters for telephony is given in Chapter 17. Even though the matrix was developed originally for the determination of customers' QoS criteria, it can be used for the determination of the offered QoS parameters to identify shortfalls and in the determination of parameters for the achieved performance in the context of specification of monitoring systems.

4.4 A methodology for the management of quality

The third component in the framework for the study and management of QoS is the methodology for management of the QoS. The management of QoS is best carried out when TQM is being practised within the service provider's organisation. TQM is described briefly in Appendix 1. Management of QoS may be carried out in isolation, but in the absence of the above practice it is unlikely to bring maximum benefit. Effective QoS management ought to be part of the overall quality culture of the service provider.

Figure 4.4 illustrates, schematically, the principal activities required for the management of QoS in an effective manner.

Step 1: The customer's QoS requirements are ascertained. This exercise may have to be carried out on a service-by-service basis and, depending on the 'granularity' of information required, customer segmentation may have to be carried out since different customers may have different QoS requirements. Determination of customer requirements is dealt with in Chapter 5.

Step 2: The customer's requirements are translated into the language usually used by the service and/or network providers as determination of customer's requirements is usually carried out using terms which are meaningful to the customers.

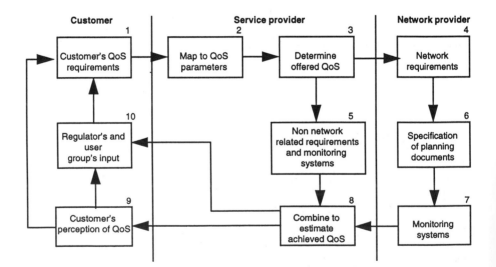

Figure 4.4   Principal activities in the study and management of QoS

Step 3: The service provider determines what level of quality is to be offered to the customers. Considerations such as cost of implementation, competitive aspects, future of the service etc. are taken into account before a level of quality is decided upon. This topic is dealt with in more detail in Chapter 6.

Steps 4–7: The offered QoS is translated into internal specifications. The network- and non-network-related aspects of quality are separated and respective planning and implementation documentation produced. Where the service provider and the network provider are the same, the documentation may be produced by the same source. Where these two entities are different, the network-related aspects are passed on to the network provider by the service provider for appropriate action. The monitoring systems for the network aspects may involve specification of test equipment, details of the frequency of measurement, sample size etc. These are dealt with in Chapter 7.

Step 8: The raw data from monitoring systems are gathered at intervals and end-to-end performance statistics produced regularly. In addition, information performance of individual elements in the network are passed on for study within the network provider. This is dealt in Chapter 7.

Steps 9 and 10: The service provider and or the Regulator or an equivalent body could carry out customer surveys to determine how satisfied the customers were with the service they had been receiving. This is dealt with in more detail in Chapter 8.

The above ten steps, when carried out, would ensure that a close watch and control were kept on the quality management. Quality is unlikely to be a major issue of contention with the service providers if customers' input is regularly obtained and the quality cycle is kept reviewed on a regular basis. The subsequent chapters in this book deal with the various aspect of this management cycle and associated activities.

4.5 Conclusions

Effective management of QoS can be made easier with the aid of the framework described in this chapter. Identification of the key issues for each of the four viewpoints gives a clear focus in their management. A clear focus on issues should enable better attention to be given for the resolution of problems. This in turn results in optimised quality, commensurate with the resources expended. The next four chapters deal in more detail the four viewpoints of QoS.

4.6 References

1   ETR 003: 'Network aspects (NAs); general aspects of quality of service (QoS) and network performance' (European Telecommunications Standard Institute, 1994), 2nd ed.
2   Richters, J. S., and Dvorak, C. A.: 'A framework for defining the quality of communications services', *IEEE Comm. Mag.,* Oct. 1988
3   'Telecommunications service quality'. US National Regulatory Research Institute, research report, March 1996.

Chapter 5

Customers' quality-of-service requirements

5.1 Introduction

The logical starting point for the study and management of QoS is the determination of customers' requirements. Customers of service providers are in both the business and residential market sectors. Not all service providers are network providers. Customers of network providers will include service providers. In this chapter, the principal characteristics of customers' QoS requirements, their applications and the management issues are dealt with.

The principal characteristics of customers' QoS requirements includes the manner in which these are expressed (on an end-to-end basis and on a

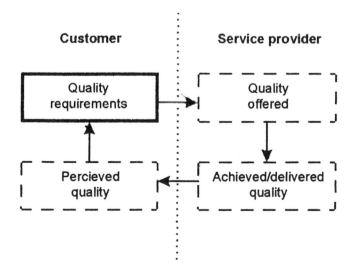

Figure 5.1   The quality cycle: customers' QoS requirements

service-by-service basis) and the factors which influence customers' requirements on quality. In the management of customers' QoS requirements, all associated principal activities are dealt with. These include derivation of service-specific QoS criteria on a service-by-service basis, methods for capturing customers' QoS requirements and how this information is subsequently submitted to the service provider's quality management process.

5.2 Characteristics

Customers' QoS requirements have the following principal characteristics:

(a) the requirements are expressed in customer understandable language and not that of the supplier;
(b) quality is expressed on an end-to-end basis;
(c) quality is expressed separately on a service-by-service basis for the principal services offered by the service provider;
(d) factors influencing requirements are:

- awareness by customers of service features and their knowledge of telecommunication technology and practice,
- relationship between price of the service and quality,
- variation of quality requirements with time,
- variation of quality levels with the level of quality experienced,
- the value of telecommunications to their lifestyle or business performance.

(a) Expression of QoS requirements in customer-understandable language

The rationale for the expression of customers' requirements in terms understandable and meaningful to the customer is stated in Section 4.2.2. For sophisticated services, e.g. services using frame relay, customers are likely to be more knowledgeable on telecommunications. Requirements are therefore likely to be expressed in sophisticated terminologies, by the telecommunications managers of businesses, in terms similar to those used by the supplier. However, the requirements of ordinary customers are more likely to be expressed in terms more meaningful to customers rather than the supplier's trade language.

(b) QoS requirements on an end-to-end basis

The customers' quality requirements are meaningfully expressed on an end-to-end basis. Customers cannot be expected to appreciate the specific characteristics of component parts of an end-end-service, but can only evaluate the service based on actual use. See also Section 4.2.2.

(c) QoS criteria on a service-by-service basis

Quality criteria could be different for different services. Even in cases where quality criteria or parameters are the same for different services, targets for these could be different for each service. For meaningful management of QoS, it is essential that each principal service has its own set of specified QoS parameters and targets.

(d) Factors influencing customers' QoS requirements

Factors influencing customers' QoS requirements may be grouped under the following headings:

(i)   type of application;
(ii) competition;
(iii) technology;
(iv) economics.

(i) *Type of application:* Figure 5.2 shows the QoS chain linking the provider and the customer. The network provider provides the network for the service provider which requires a level of QoS from the network provider. The customer served by the service provider, which could be a reseller, requires a level of QoS from the service provider. Lastly, the reseller has customers of its own and they require a level of service from the reseller. This chain of QoS requirements could result in a number of different levels of quality.

At the customer end there would be different types of customers with

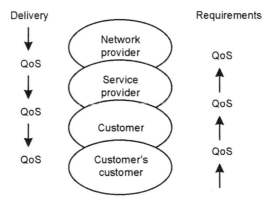

Figure 5.2   Chain of QoS requirements

specific requirements. The banking and publishing industries would require zero outage and error-free transmission during specific times of the working day. Those in continuous process industries may be able to tolerate short outages of 1 or 2 min a few times a day but not one long outage in a week. Some resellers may tolerate one long outage in a year but not many short outages in a day. The particular quality requirements for QoS parameters could similarly vary from customer to customer depending on the particular application of the service and its importance. Recognition of this factor would assist the service providers to identify the particular quality requirements of their customer base for various market sectors. Knowledge of the market and the particular customer requirements will assist in identifying the specific customer requirements on quality.

(ii) *Competition:* The bargaining power of the customers in a competitive environment will influence the level of quality they would aim for. In particular, major business customers are accordingly requiring service providers to tender against specifications which include QoS requirements. Such influences work in the telecommunications industry due to the continually improving nature of technology and the increasing communications knowledge of business customers. In cases where customers are satisfied that one performance parameter has reached an acceptable level, another parameter could take priority. Thus, quality requirements could be influenced by the improvements offered by competitors and what is realistically achievable from technological improvements.

(iii) *Technological improvements:* Customers' expectations on quality are influenced by their perception of the potential benefits of technological advances. The two principal sources of this perception are improvements in customer premises equipment and development and availability of new services based on enhanced communications platforms such as frame relay and asynchronous transfer mode (ATM). Business customers are constantly on the lookout for improved means of communication to improve the efficiency of their telecommunication applications and make them more competitive. Voice over frame relay is now possible and customers will expect this to be commercially available in due course but will expect quality of speech to be as good as with conventional digital communications.

(iv) *Economics:* The residential customer is not too concerned with high levels of quality but cost could be important. At the other extreme, large businesses whose primary activity is dependent on transactions of accurate information will require high-quality connections whenever they wish to use the service and cost will be less important. The relationship between cost (to customer) and quality is a complex one. It is an area where further research is required. Unlike retail industries, where competition has existed for a very long time and the climate for such research has existed, there has been very little investigation to understand the relationship between quality and the price a customer is willing to pay for different usage of telecommunications services.

Different market sectors require different levels of quality. The banking and publishing industries which require zero outage and error-free transmission during certain periods may pay a high price for such quality. These types of customer are, within reason, willing to pay 'any price' for quality. Other businesses, which require high quality but are cost conscious, may only be willing to pay a lower price. For such customers, while quality is

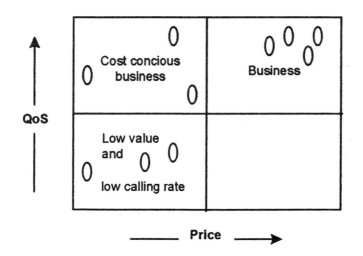

Figure 5.3   Customer population differentiated by 'value'

important, it is not vital to their survival. The residential market, which traditionally makes fewer calls, may be content with poorer quality, but is also cost conscious. Customer differentiation is shown in **Figure 5.3**.

The service provider needs to be aware of the different dimensions of customers' QoS requirements and the factors that influence these. Such awareness will differentiate the service providers in their approach to quality.

5.3 Applications of customers' QoS requirements

5.3.1 General

Customers' QoS requirements have the following applications:

(a) these form the *basis* for the determination of the level of quality offered by the service provider both to the customers at large, and in the form of SLAs with individual customers;
(b) these are the basis for the regulator to establish quality parameters on which achieved performance data should be published by service providers;
(c) these act as a guide to equipment manufacturers.

5.3.2 QoS requirements as the basis for the provision of service

These are the formal input to the quality cycle and the management of QoS by the suppliers. In the monopoly era, the supplier decided what was good enough for the customers. Those days are fast disappearing, especially in environments where competition and privatisation exist. For an effective and meaningful dialogue on quality between customers and service providers, the requirements on quality become the starting point. In a competitive environment the customer has the room to negotiate with the service providers to achieve an equilibrium between quality and price. While this dialogue may be restricted to larger business users, the principle may be equally well applied by the service provider to estimate the likely revenues against varying levels of quality for various segments of the customer population.

Service level agreements (SLAs) are usually made between the supplier and a single customer. These are contractually binding with financial penalties for noncompliance. SLAs could also exist between a network provider and a service provider and between one supplier and another supplier, as in international connections. SLAs have more than the quality element, and also cover issues on maintenance agreements, tariff matters and other issues customers consider important enough to be incorporated in some form of guarantee. The starting point for negotiations for a SLA is the customers' requirement.

5.3.3 QoS parameters for performance comparisons

Customers' performance requirements will be the basis for the regulator to specify the performance parameters on which to publish achieved results. The regulator may choose to specify these parameters on a service-by-service basis, as such publications are demanded by a significant proportion of the customers.

5.3.4 QoS requirements as a guide for equipment manufacturers

Some test equipment manufacturers take into account customers' requirements in their design of measuring equipment. Consideration of customers' quality requirements in the design of telecommunication switching and transmission equipment by equipment manufacturers is marginal, as they are usually guided by the requirement set by service providers and network providers. In these cases the manufacturers do benefit, directly or indirectly, from customers' requirements on quality.

In effective management of QoS, customers' requirements becomes the formal input to the quality cycle described in Section 4.2. Such an input will enable the supplier to ensure that the network and operational aspects are neither over-engineered nor under-engineered, but optimised for the maximum benefit of both parties.

5.4 Relevant parties in the management of customers' QoS requirements

5.4.1 General

The following parties have varying degrees of involvement in the management of quality of a telecommunication service:

- service providers;
- network providers;
- international standardisation bodies;
- user groups;
- regulators.

5.4.2 Service providers

The service provider is the principal party responsible for management of the QoS. The large service providers own and manage their networks, e.g. AT&T, BT, NTT and Telstra. The smaller service providers may lease network facilities, either in whole or in part, from a main network provider. The customers always deal with the service providers whose responsibility is to ensure that quality, whether network related or not, is of the desired level.

5.4.3 Network providers

The network provider is responsible for the network-related quality criteria delivered to the service provider. Where the network provider is not the service provider, formal agreements (SLAs) with the service provider may exist for the levels of quality. Network providers could have agreements with several service providers. They have to address the issue of managing network quality, possibly for different levels of quality demanded by the service providers.

5.4.4 International standardisation bodies

The standardisation bodies specify the technical performance of telecommunication networks and services. Most of the work within the standardisation bodies has been carried out by the representatives from large service providers which also own their networks. These standards, when incorporated into network and service management, should result in the prescribed level of end-to-end quality. To date, these specifications have been geared mainly to meet the needs of the network provider and not that of the customer. However, this situation is expected to change. It is expected that more standards to meet customers' requirements will be issued. The role of standardisation bodies is further discussed in Chapter 13.

5.4.5 User groups

User groups exercise influence by stating the performance parameters which are preferred for the publication of achieved quality. Service providers sometimes select QoS parameters for publication which give the impression of good overall performance. However, as the power of user groups increase in a competitive environment, it is quite likely that their requirements for published QoS parameters will be respected by the service providers in the future.

5.4.6 Regulators

Since the principal role of the regulator is to ensure that customers receive value for money, they have become concerned with the quality provided by the suppliers and require QoS parameters to be published on a regular (usually quarterly) basis. For example, in the UK, the Office of Telecommunications (OFTEL) requires delivered performance to be published every quarter. Even though their involvement in QoS has been to a great extent restricted to the publication by the supplier of their achieved quality, their influence has only marginally impinged on customers' requirements of quality levels.

5.5 Management of customers' QoS requirements

5.5.1 The principal issues

The principal issues to be addressed in the management of customers' QoS requirements are grouped under the following headings (see Figure 5.4):

- derivation of service-specific QoS criteria;
- capture of customers' QoS requirements;
- determination of customer groupings;
- sample size for customer survey;
- design of customer-requirements questionnaire;
- processing of data gathered;
- determination of QoS criteria for new services; and
- revision of QoS criteria for existing services.

5.5.2 Derivation of service-specific QoS criteria

Derivation of the service-specific QoS criteria and parameters may be carried out by an interactive process. The generic matrix is explored, cell by cell, for possible performance criteria, where applicable comparing the existing or known performance measures for the service in question. By this process, QoS parameters, not hitherto considered, are likely to be captured. With the help of the matrix, this may be used to capture a comprehensive range of QoS requirements. Table 5.1 illustrates a set of quantitative QoS measures for basic telephony obtained by this interactive process.

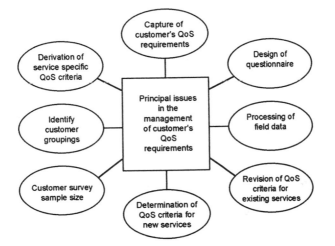

Figure 5.4   Principle issues in the management of customers' QoS requirements

Table 5.1   Quantitative QoS criteria for basic telephony derived from the matrix

Criterion/ parameter	QoS criteria or performance parameter (cell references from matrix in Figure 4.3 are given in brackets)
1	Speed of obtaining precontract information on service from provider, e.g. tariff, service availability, service features, choice of telephone features etc. (sales/speed)
2	Professionalism with which a customer is handled by the supplier (sales/reliability)
3	Speed of complaints handling (service support/speed)
4	Speed of repair (repair/speed)
5	Repairs carried out right first time (repair/accuracy)
6	Accuracy of reaching destination first time, i.e. absence of misrouting (connection set-up/accuracy)
7	Availability of network resources when requiring to make a call (connection set-up/availability)
8	Respect of wishes made with regard to display of calling-line identity (connection set-up/security)
9	Delay on making long-distance calls, i.e. call-set-up time (connection set-up/speed)
10	Transmission delay especially on international calls (information-transfer/speed)
11	Call quality (combined effect of noise, clicks, sidetone, other degrading factors (information-transfer/accuracy)
12	Availability of network resources to keep call for the intended duration of the conversation (information-transfer/availability)
13	Connection release as specified by the supplier (connection-release/accuracy)
14	Charging and billing accuracy (charging and billing/accuracy)

The granularity of the QoS criteria may be selected to suit the individual requirement of a service provider. For basic telephony, finer granularity may be achieved by specifying quality criteria based on derivations from further cells. Up to 43 have been identified for basic telephony.

5.5.3 *Methods of capture of customers' QoS requirements for established services*

Capture of customers' QoS requirements may be attempted by a combination of the following methods:

(a) questionnaires;
(b) face-to-face interviews;
(c) telephone interviews;
(d) analysis of complaints profile;
(e) case studies.

(a) Questionnaires

Questionnaires are probably the most obvious choice for the person seeking information. The respondents are asked to fill in the required level of quality by quoting figures for quantitative parameters and state in their own words the qualitative levels for other parameters. The questionnaire must state the time frame to which the quality levels will apply in the future. It must not ask how satisfied the customers are about the service currently received. This will not capture their requirements, but will capture their perception of past service.

The percentage of respondents replying to questionnaires usually varies from 1% to 50% and sometimes higher. The determining factors for the level of replies include:

- the level of loyalty felt towards the questioner;
- the level of benefit perceived by the respondent in returning a questionnaire;
- the level of interest in the content of the questionnaire; and
- the amount of time available at the disposal of the respondent.

Seeking information by questionnaires has limitations. It is reasonable to assume that more questionnaires will be completed by people who have a 'complaint' to make. The results of a survey may, therefore, automatically introduce a bias into the estimation of the requirement of the whole population. This argument, while technically plausible, is not of great concern to this exercise as the people who make complaints are those who form opinions about the quality of a service. The people who have complaints to make are also, generally, those who place exacting demands on the service. These users mostly also happen to be large business users and therefore their opinions cannot be directly equated to those of residential customers.

In filling questionnaires, respondents do not usually have the means of clarifying questions which they may not understand. This could result in vague answers. Some parts of the questionnaires may be left blank. There may also be some questions which the respondents may have misunderstood. Notwithstanding these shortcomings, questionnaires are a powerful tool to capture customers' requirements. Many of the above shortcomings can be overcome by increasing the sample size and by intelligent analysis of the respondents' replies.

(b) Face-to-face interviews

Face-to-face interviews, with a questionnaire, should provide the ideal alternative to the postal questionnaire. The shortcomings mentioned in the previous method can be eliminated. However, face-to-face interviews are time consuming and make a high demand on resources. Nevertheless, it is always useful to have some face-to-face interviews, if only for the analyst to acquire an appreciation of the customers' problems and concerns which are not covered in the questionnaire. This should enable an intelligent interpretation to be made of the replies on questionnaires and may also help with interpretation of postal questionnaires. Face-to-face interviews are particularly useful for the capture of qualitative QoS criteria.

(c) Telephone interviews

Telephone questionnaires are a compromise between postal questionnaire returns and face-to-face interviews. The principal shortcoming of this type of interview is the absence of body language of the respondent. This limits the understanding by the questioner of the respondent's concerns. This method is, however, better than postal-questionnaire returns without any form of person-to-person contact.

Where possible, a combination of postal questionnaires, telephone interviews and face-to-face discussions should be carried out to achieve a good representation of customers' QoS requirements.

(d) Analysis of complaints profile

Analysis of complaints, particularly related to the service in question, is a source of information on what customers expect in terms of quality. A customer usually complains when dissatisfaction has reached an unacceptable level. By studying the pattern, the analyst can ascertain the QoS aspects which are of most concern to the customers. This source should not be considered as the main one for the capture of customers' QoS requirements, but only to supplement the information gathered by the means described above.

(e) Case studies

Ascertaining customers' QoS requirements from case studies may be resorted to whenever opportunity arises. In many first-hand experiences of customers, based on multiple observations of various tasks carried out by the suppliers, customers are able to provide valuable insights into the weaknesses of the service providers and what level of quality they normally expect.

5.5.4 Determination of customer groupings

For a given telecommunication service, different customers can have different requirements. Groups of customers with similar requirements, i.e. market segments, can be identified and their common requirements, where such groups exist, may then be obtained. For example, hospitals, the retail trade, airline booking offices etc. have their own unique QoS requirements. For hospitals it is availability round the clock throughout the year. For the retail trade, it is availability during trading hours. For airline booking offices, availability of network resources and fast response times are critical.

The conventional way of grouping according to market sector and segment is the preferred method. The population is first divided according to sectors, e.g. banking, electrical, transport etc. The Standard Industrial Classification [1] may be used as a starting point and modified in the light of known data for the population under study. Then segmentation may be carried out. For example segmentation into multinational, national and local segments of each sector may be necessary. These segments could have differing telecommunication requirements, In the light of experience these groupings may be revised to suit the particular characteristics of a service for a particular country.

These procedures are rigorous, but short cuts should only be taken when warranted by evidence that certain procedures may be eliminated. If short cuts normally taken, this practice must be weighted against the risk of missing unique customer requirements, which may prove costly to rectify at a later stage.

5.5.5 Sample size

The sample size for administering the questionnaire to ascertain QoS requirements for a particular group of customers is dependent on the level of accuracy required. The accuracy may depend on the nature of the individual samples and the distribution of the samples in the population. If the individual samples are considered to be members of a normal distribution, the selection task is straightforward. It will be necessary for the service provider to carry out some basic research and analysis of the sample to establish if it is representative of the group. In the absence of any further information, the population's requirements may be assumed to follow a normal distribution. The formula for the range of mean of the total population estimated from the sample mean for given levels of confidence are found in standard textbooks on statistics. Mathematical treatment of sampling theory is not within the scope of this book. Standard textbooks on statistics should provide guidance in this area [2].

In practice, the judgement is also likely to be based on both a qualitative basis and a mathematical basis. It will be up to the service provider to assess, from the sample, whether it represents a consensus view of the group. This is probably the better way of estimating requirements of the group than a mathematical method which excludes the subjective element. The subjective element, when it is an informed one, can be very meaningful.

5.5.6 Design of questionnaire

The following guidelines should be borne in mind in the design of a questionnaire to capture customers' QoS requirements:

(a) The questionnaire should be designed around the QoS criteria derived for the service from the framework, but there must be sufficient flexibility to accommodate customers' answers expressed in their own words. These could later be translated to terms used within the service providers' organisation.

(b) Questions must be worded in a manner which reflects the customers' cultural thinking on the quality of a particular service. In other words, the questioner should relate to the respondent. In the absence of a person readily available to clarify enquiries, the questions must be unambiguous and clearly impress on the respondent's mind exactly what is being asked for.

(c) The questionnaire should also be geared to obtain the qualitative issues from customers.

The questionnaire should be designed to find out what the customers' QoS requirements are and not what the service provider thinks the customers' requirements should be. This is most important in a competitive environment.

Design of questionnaires is a specialised subject [3, 4], and no attempt is given in this book to dwell on this in any detail. Telecommunications engineers must resist the temptation to dismiss the topic of questionnaire design as a task requiring a low level of intellectual skills. Design of a good questionnaire requires an understanding of psychology, cultural behaviour of the respondents, customers' knowledge of the telecommunications services, quality implications, dynamics of market forces and the customers' expectations, and an awareness of competitors' products and services. Service providers, network providers, regulators and user groups should note that, with experience and professional advice, meaningful information can be obtained from adequately designed questionnaires. A poorly designed questionnaire could produce misleading results and wastes customers' and the questioner's resources.

Four categories of responses may be captured from the customers relating to their QoS requirements. These are:

(a) levels of quality for the quantitative performance parameters;
(b) narrative requirements for the qualitative performance parameters;
(c) priority of these parameters; and
(d) any other quality issue relevant to the study, but not included in the questionnaire.

Levels of quality for each parameter may also be expressed qualitatively by the customer. Customers conversant with telecommunication terms (telecommunication managers of large companies) may be able to supply quantitative levels. It is possible that some concerns may not be identified by the questionnaire. To capture these, customers may be asked to state any quality issues they wish the service provider to address but which are not covered in the questionnaire.

The most effective method of administering a questionnaire for capturing customers' QoS requirements is to interview the customer personally. However, it may be more economical to use this approach with large business customers and, to a lesser degree, with residential customers and those with special requirements, e.g. the blind and slightly hard of hearing.

Although the practice of obtaining customers' requirement by means of questionnaires is popular, this method of obtaining results should be restricted to the minimum, as maximum understanding of customers' requirements is obtained by personal interactive association with customers.

The task of ascertaining customers' QoS requirements is best given to a market research agency which is competent in telecommunications. A pattern will soon emerge after interviews of a small number of customers within the same group; thus after a certain number of interviews, further interviews may not be considered necessary. With interaction between the service provider and the agency, it should be possible to obtain a meaningful representation of the customers' requirements. In the design and administration of the questionnaire, it is essential for the service provider to be familiar with the cultural aspects of the customers. This will provide a better understanding of the answers given by the customer which could result in a more credible representation of the customers' requirements.

A typical question from a questionnaire on a quantitative measure is illustrated in Figure 5.5.

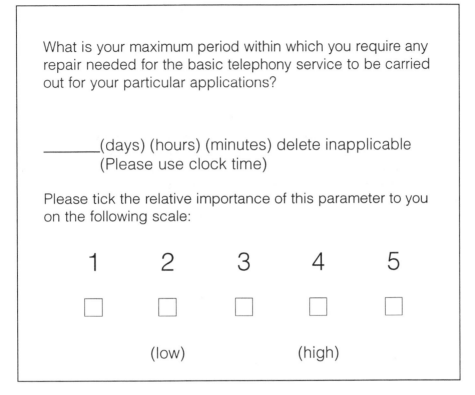

What is your maximum period within which you require any repair needed for the basic telephony service to be carried out for your particular applications?

_____(days) (hours) (minutes) delete inapplicable
(Please use clock time)

Please tick the relative importance of this parameter to you on the following scale:

1   2   3   4   5

☐   ☐   ☐   ☐   ☐

(low)   (high)

Figure 5.5   Question on the maximum repair time

5.5.7 Processing of customer responses

The quantitative answers given for each parameter are processed to estimate the customers' requirements. Table 5.2 shows how this may be carried out.

Table 5.2   Customers' preferred maximum time for repair

Maximum time for repair (h)	1	1.5	2	2.5	3	3.5	4
Number of customers	10	7	7	7	5	3	1
Percentage of population	25	17.5	17.5	17.5	12.5	7.5	2.5
Cumulative percentage	100	75.0	57.5	40.0	22.5	10.0	2.5

The distribution may be represented by the histogram in Figure 5.6.

The cumulative percentage of the population of the sample may be plotted against the maximum repair time and is shown in Figure 5.7. It is seen that to satisfy 90% of the customers, the service provider has to arrange for a maximum repair time of around one and a quarter hours.

This type of analysis may be carried out, on a parameter-by-parameter basis, from the replies received from customers for submission into the next stage of the quality cycle, the determination of the offered QoS.

In the treatment of customers' requirements, it must be appreciated that conclusions obtained from statistical information should not be treated in a clinical manner. Management of QoS is both a science and an art. Familiarity with customers' requirements, their past responses, if any, and industry knowledge will enable the analyst to conclude that any spurious, inconsistent or unreliable answers should be ignored. Meaningful customers' requirements can only be obtained by the experienced analyst, and this is as much an art as a science. The contributory factors for this acquired interpretative knowledge are: familiarity with quality parameters, the currently achieved level of performance for these parameters, how these will affect the customers' attitude towards the supplier, what the competition offers, what level of performance is actually achievable etc. 'Reading between the lines' to assess what customers really expect is of better value than mathematical analysis of the figures.

From the analysis of the responses on a parameter-by-parameter basis, a set of figures may be estimated to represent customers' level of performance required. For criteria where qualitative narrations are given by the customer, these are analysed to estimate the required performance. For example, with the following set of narrations from different customers, it is clear what the general message is.

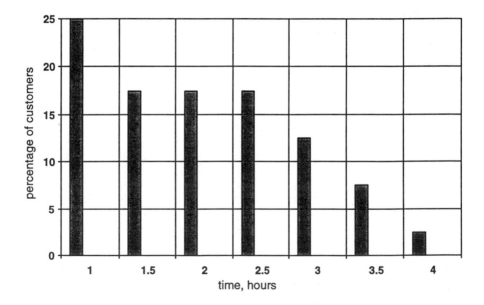

Figure 5.6   Customers' maximum repair time distribution

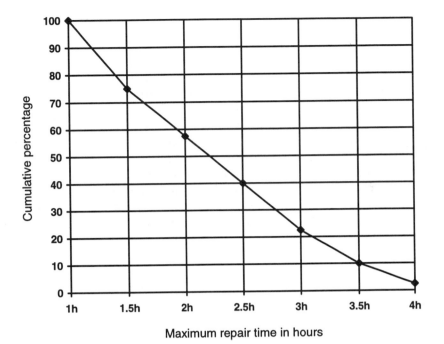

Figure 5.7 Cumulative percentage of customers and maximum repair time

Examples of narrations:

(i)   billing details should be simple and clear;
(ii)   more consolidation – less bills;
(iii)   more information on nonstandard billing entries (one-off charges);
(iv)   details of each line and one bill;
(v)   details could be more thorough;
(vi)   presentation could be more thorough;
(vii)   one bill format OK;
(viii) should include our cost centres;
(ix)   bill every month as opposed to currently every two months;
(x)   default payments by service provider for not meeting SLA to be indicated.

The conclusion is that customers require customisation of bills on an individually tailored basis.

Integration of customer responses

It is possible that different customer groupings have different needs for the same parameter. For example, availability would be crucially important 24 hours a day to the hospital sector, but only important to the retail industry over part of the day. Industry-specific requirements should be identified from the requirements and highlighted in the input to the quality cycle. Unique industry requirements should be treated separately to give these the attention they require in the quality cycle. Combination of parameters may be attempted should this produce no conflict of benefit to the customers.

Prioritisation of QoS requirements

In the example questionnaire in Section 5.5.6 (Figure 5.5), the customer was asked to rate the importance of a parameter. With an adequate number of replies, the mean of such responses would give a fairly accurate indication of the importance of this parameter. With the computation of mean rating of the 14 parameters for basic telephony (as in Table 5.1), the parameters may be ranked (in the order of mean-rating scores), resulting in a prioritised list. A precise mathematical interpretation ought not to be taken too seriously; however, this method of prioritising produces a fairly reliable guide to how customers rate a particular parameter.

Editing of customer requirements to terms used by service providers

Certain QoS criteria will need to be translated into the service provider's language. For example, if customers are asked to state for how many calls in a 100 they are prepared to tolerate moderate difficulty in understanding during the conversation phase, this has to be translated into meaningful technical parameters for the service provider to act on.

Algorithms exist for the percentage difficulty customers are likely to experience with degrading parameters such as sidetone, echo and noise. Limits of performance of the network may be derived from these, should the service provider choose to meet the customers' requirements. Similarly, availability may sometimes be expressed in both qualitative and quantitative terms by the customer. This ought to be expressed mathematically for the service provider to act on.

Conversion of customers' requirements into service provider's language entails familiarity with the service provider's terminologies, and is to be attempted by the service provider rather than a nontelecommunication professional. The information is now ready for input to the next stage of the quality cycle.

5.5.8 New services

Capture of customers' QoS requirements for new services is complicated by the fact that no experience exists for the information seeker to draw on. Customers may, at best, have only a sketchy idea of the new service and this may not be adequate for them to provide meaningful information on required levels of quality. The following guidelines should assist in the determination of the QoS parameters of any new service:

(i) The service provider could, where possible, break down the service elements, as seen by the customers, into similar service features in existing telecommunication services. For example, in the determination of QoS parameters for multimedia communications, the following components may be identified as those which exist on other telecommunication services:

- video screens on computers;
- data transmission;
- network access and retrieval;
- data security.

It is possible to identify the QoS parameters for each of these service features, and then attempt to integrate these for the new multimedia service.

(ii) Having formulated a tentative list of parameters, a dialogue could be established with potential users and the list modified where necessary.

(iii) A trial of the service based on the tentative QoS parameters may provide feedback, enabling a more accurate set of parameters to be identified.

Further refining should enable a set of relevant QoS parameters to be obtained for a new service. Such an exercise should commence at the beginning of the launch of a new service. In the absence of such an exercise, it may be difficult to make corrections, if intractable service features or installation procedures have been built into the service.

It is essential, when designing new services, to include in the design specification the appropriate technical performance to give an adequate QoS to users of the service. Both population and usage modelling should be carried out to ensure that QoS is not jeopardised by unusual usage patterns or feature interaction with existing services.

5.5.9 Revision of QoS requirements

The quality requirements of customers inevitably become more demanding with time; when their expectations are satisfied they look for improvements — this is a general feature for all customers. It therefore becomes necessary for the service provider to:

(i) identify when revisions are required;
(ii) identify the nature of the changes; and
(iii) ascertain the necessary changes in priorities.

Intimate knowledge of the customers' needs, industry movements, competitive threats and technological developments will assist the service provider in the identification of these aspects.

5.6 Customers' QoS requirements and the quality cycle

The processed customers' requirements are now ready to be fed into the next stage of the quality cycle, which is the offered QoS. When varying levels of performance are required for a certain parameter, a dilemma is posed to the service provider. Notwithstanding the different levels of quality requirement, the consolidated quality requirements are submitted to the decision-making process for determination of the level of quality the service provider plans to offer.

5.7 Conclusions

Management of customers' QoS requirements has its own unique set of issues to be addressed. These are best identified and focused by decoupling management issues from other viewpoints of QoS. Service-specific QoS parameters are first derived for the service under study, then a workable method of capturing these requirements is developed. The captured information, after processing, is ready for use in the next stage of the quality cycle, the offered QoS.

5.8 References

1   CENTRAL STATISTICAL OFFICE: 'Standard Industrial Classification' (HMSO, London)
2   CURWIN, J., and SLATER, A.: 'Quantitative methods for business decisions' (Chapman & Hall, 1994)
3   SUDMAN, S., and BRADBURN, N.: 'Asking questions: a practical guide to questionnaire design' (Jossey-Bass, San Francisco, 1983)
4   OPPENHEIM, A. N.: 'Questionnaire design, interviewing and attitude measurement' (Printer Publishers, 1996)

Chapter 6

Quality of service offered

6.1 Introduction

The service providers' view of quality has traditionally been known as the planned or targeted level of service quality. If a formal specification of offered quality to the customer existed, it would form part of the internal performance specification of a telecommunications service and also be the basis of the service provider's quality measures. However, despite the fact that telecommunications can trace its origins through telegraphy back to the middle of the 19th century, very few telecommunications service providers offer formal specifications of quality for their services. The exception would be the service level agreements (SLAs) to a few customers. A quality specification should be a subset of the normal service specification.

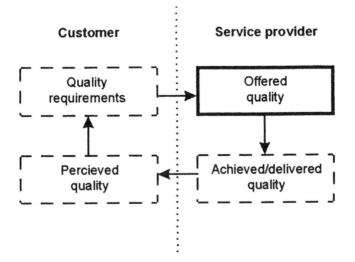

Figure 6.1   Quality cycle: offered quality

The major reasons for the lack of formal quality specifications lie in the history and development of the telecommunications industry to date. Telecommunications-service provision was traditionally offered by government-owned technical organisations operating in monopolistic environments. The main business objective pursued by these organisations was to 'provide a service to its subscribers'. Today the term 'subscriber' has been replaced with the term 'customer'. The main strategy for doing this was to divide the total communications resources available, as fairly as possible, among all the customers irrespective of need. This was a broad-brush approach, but telecommunication organisations were monopolies and therefore could decide what happened within the industry without undue pressure from outside. Therefore, there was very little incentive to specify customer-oriented service performance or quality.

With the onset of competition, privatisation and regulation in different parts of the world, telecommunications is becoming a battle field in which service providers attempt to survive. Survival will be based around differentiation and quality is fast becoming one of the principal vehicles for differentiation, particularly for the historically incumbent operator which cannot easily compete on price because of its overheads and universal service obligations. This has resulted in the need for a formal specification of quality for each and every telecommunications service. Therefore, quality will in future form part of every service specification, along with other components such as the service description, service features, tariff, availability, service support and SLAs. This quality specification will become important to customers, regulators, competitors and any other national and international service providers. This chapter looks at the characteristics of the offered quality, its applications and its practical management.

6.2 Characteristics of offered quality

The principal characteristics of offered quality are:

(a) it is usually in the language of the telecommunications service and network provider and not the customers;
(b) it is separately specified on a service-by-service basis;
(c) there could be different levels of performance for each service; and
(d) it has a relationship with customers' quality requirements.

(a) Supplier language

Terminologies exist within the organisation of the service and network provider to express the performance and features of a service. These terms have evolved over the years and, even though there is a commonality among many service providers, there is no agreed set of definitions for the expression of QoS or service performance. Today, global and multinational customers are faced with the task of interpreting various forms of service offerings from different service providers. It is therefore impossible for customers to compare service offerings or calculate the true end-to-end service performance and compare it with their requirements. Therefore, it is vital that an internationally agreed set of service-quality definitions be agreed (not including targets of performance). This is proposed in Chapter 19.

(b) Offered quality is expressed on a service-by-service basis

For the quality to be meaningful, the performance levels of each service should be specified individually, because customers usually buy a service and not a portfolio of services. To sell a service, the quality level must be quoted so that customers know what is being bought and they can compare the service offerings from a number of service providers. Different services could also have their own unique performance parameters. Additionally, the various parameters of quality can have a different impact on different services. For example, error rates will have a major effect on data services but voice service can stand relatively high levels of error with no apparent effect on listeners. The reverse would apply to the parameter 'delay'.

(c) Different levels of performance for the same service

Because different customer groups have different requirements for the levels of performance, it is necessary for service providers to aim for correspondingly different service levels for the same service, for different customers. Therefore, service levels need not necessarily be the same for all customers; they are more likely to be tailored to meet individual requirements. Variations in performance can include not only network performance (e.g. high-quality circuits against normal-quality circuits) but different levels of service support and other non-network-related parameters. However, it is extremely difficult for a service provider to differentiate levels of service for network-related parameters, especially if the services are provided on a ubiquitous network. This may ease for future variable-bit-rate broadband networks based on asynchronous transfer mode (ATM) technology where, for example, priority on delay and bandwidth can be given to individual customers. However, it is relatively easy for a service provider to offer different levels of quality for non-network-related performance parameters to different customers for the same service.

(d) Relationship with customers' requirements

The offered quality must have a relationship with the customers' requirements in the form of an audit trail back to the original customer requirement. Where there is an absence of such a relationship, it is evident that the service provider has decided the quality level for the customers, probably based on its capability, or what it can afford, and not based on customer requirements. The relationship with the customer requirements has two features:

(i)   The offered quality must be *derived* from the customers' requirements in so far as every effort is made to meet as much of the customers' requirements as possible. The relationship would be the degree or percentage to which customer requirements were met.

(ii) The parameters with which the offered quality is *described* would have a similarity with the parameters expressed by the customers. The correspondence may be direct or indirect. For example, the customers may express the call quality qualitatively and the service provider may express it in technical parameters to achieve the customers' requirements, or in a corresponding manner.

6.3 Applications of offered quality

The principal applications of the offered quality are illustrated conceptually in Figure 6.2.

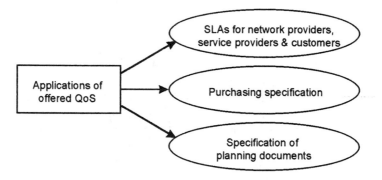

Figure 6.2   Applications of offered quality

(a) Service-level agreements

One of the principal applications of offered quality is in the determination of SLAs between the service provider and the customer or network provider dictating specific service levels. This subject is dealt with in more detail in Section 6.5.1.

(b) Information for the equipment manufacturers

The offered service performance will be both a driver for, and dependent on, the performance of the equipment needed to provide that service. The equipment service-performance specification will be a subset of the service-quality specification and will provide the manufacturer with information necessary to design and build the equipment conforming to the desired parameters and their associated targets. For example, mean time between failure (MTBF) will affect the availability of a service offered to a customer and this will depend on the component count and design of specific items of equipment.

(c) Planning within the telecommunications company

In practice, the information contained within the 'offered quality' will be used by the network provider and service provider in the planning and implementation of a service.

The offered quality forms the basis for:

- operations management, including items such as speed of provision, speed of repair, complaint-resolution time, targets and costs etc.;
- design and dimensioning of the network;
- quantifying the performance of network elements; and
- establishing monitoring systems to measure achieved-quality levels.

6.4 Parameters to express offered quality

The number of parameters specified by the service provider will usually exceed the number of parameters specified by customers. Customers are unlikely to quote all possible quality parameters for a service; they concentrate on those which are important to them and these will differ from customer to customer and from service to service. However, to make effective use of its resources, it is operationally necessary for the service provider to specify all, or as many as possible of the parameters, to satisfy the majority of its customers and then to specify variants for specific customers.

There is no internationally agreed set of parameters for the expression of 'offered quality'. For this reason, each service provider will propose its own set of quality parameters on which to specify network- and non-network-related performance criteria. The only exceptions to this will be the technical, quality-related performance parameters specified by recommendations from standards bodies such as the ITU-T and ETSI.

An illustration showing customers' quality requirements and the corresponding service providers' offered quality is shown in Table 6.1.

Table 6.1   Examples of customers' quality requirements and the corresponding offered quality parameters for basic telephony

Customers' quality requirements	Service provider offered quality
1 Specified time to obtain precontract information on a service (i.e. tariffs, availability, features, options, offers etc.).	Target time to provide the precontract information.
2 Service provider to act in a professional manner at all times.	Target number for customer-satisfaction criteria which cover professionalism and are measured on a regular basis.
3 Specified time for the resolution of complaints.	Target time for resolution of customer complaints (e.g. 90% of all complaints to be resolved in 4 hours).
4 Specified time to repair the service.	Target time for service repair (e.g. 90% of customers' reported faults to be repaired in 4 hours).
5 Faults to be repaired first time (i.e. repairs carried out right first time).	Target 90% of repairs to be completed correctly first time.
6 Accuracy of reaching destination first time, i.e. absence of misrouting	Probability of misrouting expressed as a percentage.
7 Availability of the service for use by the customer.	Target number of outages and the maximum duration of any outage.
8 Calling-line-identity requirements (e.g. nondisplay of CLI).	A number of CLI display options.
9 Delay limits on making long-distance calls (i.e. call-set-up time).	Maximum target call-set-up time for international calls.
10 Transmission-delay requirements, especially on international connections.	Target delays for different services expressed in milliseconds.
11 Call-quality requirements.	Targets on performance parameters which contribute to call quality, such as loss, noise (different types), sidetone, echo, error rates etc.
12 Service availability for the intended duration of use.	Availability of end-to-end service for the intended duration of use expressed as a percentage, total number of outages and duration of longest outage.
13 Flexibility in the provision of a service.	Options to customise outside of the normal options.
14 Charging and billing accuracy	Charging and billing accuracy specified in terms of: (i) maximum number of errors, and (ii) maximum magnitude of any error.

6.5 Management of offered quality

The principal areas to be addressed in the management of offered quality
are:

(a) determination of the 'offered quality';
(b) specification in planning documents for use by the service and network
provider; and
(c) specification of monitoring and measuring systems.

The first two are dealt with in this chapter and monitoring management
systems are explained in Chapter 7.

6.5.1 Determination of offered quality

The service and network providers consider customer requirements along
with other internal and external business considerations in order to
determine the optimum practical quality levels. The framework shown in
Figure 6.3 illustrates how the offered quality may be determined.

Figure 6.3 shows the process of determining the offered quality from
customers' requirements. There should be a dialogue between the customers
and the service provider during the service provider's internal capability
assessment, in order that a number of options with various quality levels and
costs can be considered. In practice, this will normally take place between the
service provider and a few large customers representing the market sector at
which the service is directed. On completion of the internal capability
assessment, the service provider will be in a position to decide on the level of
offered quality. This may not meet all the original customer requirements,
but it will be the most appropriate fit between customer requirements, cost
and the service providers' capabilities.

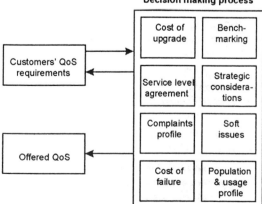

Figure 6.3   Process for the determination of offered quality

The principal considerations in arriving at the offered quality may be classified under the following headings:

(a) cost of providing the required quality (e.g. by upgrading);
(b) benchmarking;
(c) SLAs;
(d) strategic considerations (time scales, competition, regulatory require-
     ments, product life cycle etc.);
(e) cost of failure to meet offered quality (e.g. revenue rebates);
(f) population and usage criteria;
(g) complaints; and
(h) qualitative criteria (i.e. 'soft' issues).

The first six are identified and discussed in this chapter and the remainder are discussed in Chapter 9.

(a) Cost of quality improvements

To satisfy many customer requirements, enhancements to quality may be required. In the unlikely event of offered quality being reduced, there must be an understanding of the potential penalties which may accrue, e.g. loss of market share. With the complex service-related cost structures available to most service and network providers, it will be difficult to produce an estimate of the relationship between quality, cost and financial benefits. However, it is sometimes possible to arrive at a relationship between increments in quality and cost of improvement. In the process of determining offered quality, the incremental cost will be one of the inputs. The opportunity cost of not improving quality and the possible resulting loss of revenue also have to be addressed. The network and service providers have a responsibility to develop a quality-versus-cost relationship in the future to obtain fair pricing of quality to their customers. (This is further discussed in Chapter 12.)

(b) Benchmarking

Benchmarking is the comparison of a company's performance with that of competitors and like companies [1]. Often it results in a league table of parameters where the performance of a company is compared with that of others where individual parameters are considered best, e.g. fastest in provision of repair of service, best customer satisfaction, lowest cost, highest connection per employee etc. Care must be taken to ensure that there is a valid comparison between service providers for each parameter, e.g. a service provider with a lowest connection per employee may provide both local and trunk service, whereas one with highest connections per employee may only provide a local service. Once the benchmarks have been established, the service provider can set in motion activities to bridge the gaps between its performance and the 'best of breed'. The model for carrying out benchmarking is shown in Figure 0.4.

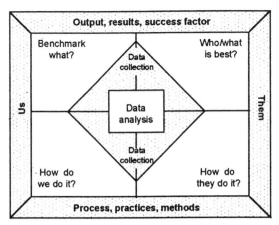

Figure 6.4   Rank Xerox benchmarking model [2]

The use of 'snake charts', illustrated in Figure 6.5, is a useful way of comparing performance against that of a competitor graphically and highlighting areas for improvement.

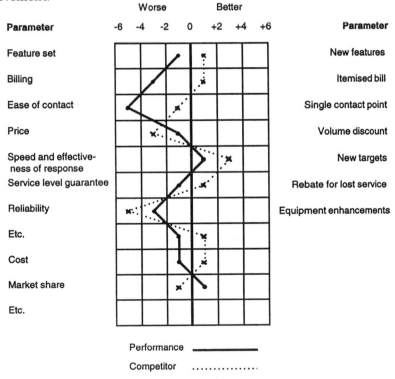

Figure 6.5   'Snake chart' illustrating benchmarking

An alternative pictorial way of comparing the performance of a service with that of a competitor is shown in Figure 6.6.

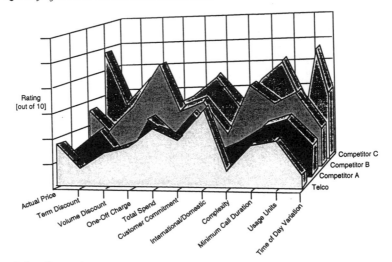

Rating
[out of 10]

Competitor C
Competitor B
Competitor A
Telco

Actual Price
Term Discount
Volume Discount
One-Off Charge
Total Spend
Customer Commitment
International/Domestic
Complexity
Minimum Call Duration
Usage Units
Time of Day Variation

Figure 6.6   Competitor positioning

Benchmarking is also about improvement through change [2]. A report of recent research states that, of the 1000 companies listed in *The Times,* 67% claim to have used benchmarking and 82% of these used it successfully. The usefulness of benchmarking has been well recognised by the more successful companies. Telecommunications-service providers are no exception; with increasing pressure to perform better with reduced resources, benchmarking is more widely becoming an input to the business-planning and performance-assessment processes. Another dimension of benchmarking is adoption of the best practices of another organisation. This best practice need not be from the same industry; it may be translated to apply to the needs of another industry.

To benchmark, similar service-performance statistics must be examined on a comparable basis. In practice three principal difficulties exist. The service features are not always similar, performance definitions may differ and performance statistics do not conform to any commonly agreed methods of capture and presentation. True comparisons of performance become a combination of art and science. A comparative framework has to be developed by the individual service and network provider to give relevant weighting for various issues, such as market share, purchasing power in the local currency, service-usage habits and so on, to enable a valid comparison to be made.

Another issue related to benchmarking is that it is not a constant performance figure, because most service and network providers aim to improve the quality of their services over time. If a service provider wishes to becomes the top service and network provider for quality, then it must outdo other competitors in the long term.

Another issue to be addressed is the question of ethics. Benchmarking needs to be practised without recourse to industrial espionage. Information on performance may be closely guarded by service and network providers in the interests of competition. It will be a test of the integrity of a service or network provider to provide a level of service the customer wants and to differentiate itself from its competitors without breaking the normally accepted ethical practices in the industry. A useful reference on benchmarking ethics is 'The benchmarking code of conduct' developed jointly by the American Productivity and Quality Centre's Clearinghouse and the Strategic Planning Institute (SPI) Council on Benchmarking [2].

(c) Service-level agreements

SLAs are useful for the suppliers to formalise (often contractual) relationships with:

(i)   other network providers abroad;
(ii)   other network and service providers in the same country with which interconnection exists;
(iii) individual users; and
(iv)   business customers.

Increasingly, with more complex network interactions and sophisticated internetworked services, agreement must also cover the possibility of one network compromising the integrity of another. (This aspect is covered in Chapter 11.) The principal management issues for each of the above categories are examined below.

(i) *SLAs with other service and network providers abroad:* As telecommunications services are provided on a global basis, it is necessary for a service or network provider to interface with service or network providers within other countries, either to provide a service to that country or to transit a service to a third country. Therefore, services can span more than one country and more than one foreign service or network provider.

There are two principal types of issues to be addressed in dealing with quality on a global and end-to-end basis: interworking (compatibility) and the level of service quality which the two parties to the SLA plan to offer to their customers.

Interworking between two networks: A network in one country connected to a network in another country has to conform to certain standards in order that they interwork. Recommendations have been formulated in ITU-T and other standards bodies for all the essential interface requirements. What is lacking is a comprehensive list of all the end-to-end quality-of-service parameters and associated technical-performance figures, together with apportionment values between networks, in order that the individual service and network providers may design for their respective performance figures. Certain performance figures exist, e.g. transmission delay in ITU-T *G.114* [3]. Even though some of the major parameters are covered, there is a need to extend these to specify end-to-end performance and associated quality for a range of parameters for all the principal services. Chapter 19 makes out a case for this and proposes a framework for its study.

Mutually preferred end-to-end performance between two administrations: In some cases, a service/network provider may wish to liaise with a service/network provider from another country to improve the end-to-end performance of services between these two providers. This presents problems for both service or network providers. The first is to find another service provider with the best service fit, in terms of price and functionality. The second, is to establish the desired quality levels which will form the contents of an agreed SLA between the two suppliers.

This is a difficult task mainly because of the lack of internationally agreed definitions covering quality criteria. For an improvement in quality, the service providers will first need to agree on a common set of quality definitions for the required quality criteria needing improvement. For example, one service provider may quote 'time to repair' as 4 hours, with the definition that this covers 'the instant the problem is reported by the customer, to the instant the problem has been resolved to the customer's satisfaction'. The other service provider may quote 'time to repair' as 3 hours, with the definition that this covers 'the instant that the problem has been reported to its operational staff to the instant the problem has been rectified to the customer's satisfaction'. However, if the elapsed time between the customer reporting the problem and the operating staff receiving notification of the problem is 3 hours, then the real 'time to repair' from the customers perspective is 6 hours and not 3 hours as quoted by this operator. Therefore, achievement of improved quality levels dictated by market forces prompts the participating service providers to work out formal SLAs between themselves with agreed definitions to meet the desired level of quality.

In arriving at a SLA between two service providers, the following considerations must be addressed. Cultural differences may influence the level of service required in each country. For example, the customer quality requirements for 'time to repair' a telephone service in the UK could be 4 hours and in France this could be 6 hours for the same service. However, this service may span both countries for the same (i.e. the multinational) customer. Therefore, a company having offices in the UK and France will be quoted two different repair times from two suppliers for the same service.

The vision and objectives of each service provider may differ. A service provider may be content to provide the basic level of service to its customers. This may be due to lack of competition, the service providers' resource capability and the maturity of the market (i.e. market penetration). For example, in a developing country where the market penetration is very low, the service provider's priority could be to provide service and increase penetration at the expense of quality. In this case, it will be very difficult for another service provider to agree an SLA covering tighter quality criteria for a service interconnecting both service providers' customers. This may result in both service providers agreeing an SLA with the minimum quality level acceptable within the ITU recommendations.

For successful management of SLAs between two service providers from different countries, it is necessary to specify quality levels using the same performance parameters. Agreements must be reached between the two service providers on a common understanding of performance parameters. Where different methods of measurements and specification exist, for example in speech-loudness levels, agreement ought to be pursued on what constitutes mutually acceptable performance definitions and how these may be translated into understandable and meaningful levels of quality by both service providers. The service providers should then agree an SLA, in writing, covering the levels of performance together with their definitions and should also specify their respective roles and responsibilities in the operation of the SLA.

(ii) *SLAs with other national network or service providers:* Service providers within a country may need to use each other's capability in order to reach the total customer base within that country. This is commonly known as 'interconnect'. The end-to-end performance will be dependent on service providers' capabilities and, again, the quality offered to the customer will be based around an agreed SLA between the two service providers commensurate with their combined quality capability. In managing this type of SLA, the following points should be taken into account:

- the service providers concerned should agree, preferably contractually, the parameters and their performance at their interface;
- they should specify these on a service-by-service basis; and
- they should establish and agree the procedures in the event of non-compliance with the SLAs.

In some countries, the regulator provides the guidelines within which the service providers should interwork, and arbitrates in the case of disputes.

(iii) *Individual customers:* Residential or personal customers are not very discerning and generally accept a fairly basic level of quality in relation to price. This fact is exploited by service providers, since price is often the major issue and few customers will incur the inconvenience of changing providers on the basis of a nominal price differential. Any such move would be opportunistic, such as during a house move, or be prompted by emotional reaction to an unsatisfactory experience.

(iv) *Business customers:* At the other end of the spectrum (i.e. medium and large business customers) quality is a much larger issue, for example, near 100% reliability is required by customers such as banks, airline companies and those transferring financial data. For these types of customer, reliability is of utmost importance; consequently they will pay a premium price for high levels of quality. SLAs will usually be individually tailored to meet their requirements.

In a liberalised environment and an environment of ever-increasing litigation, significant financial damages may be awarded against a service provider if a major customer, which is reliant on a service, sues the service provider for loss of earnings due to poor quality (e.g. a major outage). In a competitive environment, service providers often have SLAs with their major customers which are contractually binding and with large financial penalties for noncompliance. However, customers do not usually see these as ways to obtain rebates, but more as a measure of the service providers' confidence in the quality of their services. Customers do not normally wish to invoke financial compensation for breach of an SLA, but use breaches of an SLA to assess their choice of supplier (i.e. if an SLA is constantly breached are they with the correct service provider?). However, service providers must carefully consider all penalties (including financial ones) when agreeing on an SLA. For a service provider, poor performance can represent a significant loss of revenue by either direct payment or loss of a customer.

(d) Strategic considerations

In the determination of quality to be offered the service provider should consider strategic business issues, such as:

- product life cycle of the service;
- effect of competition; and
- effect on market share.

If a service is reaching the end of its useful life, as in the case of Telex, it would be not worthwhile to improve its quality. However, for new services, quality is likely to be a major contributor to market share in a competitive environment. For new entrants to the market, quality will be the prime differentiator, because the only way to increase market share is to offer equal functionality, be competitive on price and differentiate with quality. A new entrant may not wish to start a price war with the entrenched service providers, so the only viable differentiator is quality. The relationship between quality and market share will be an important input to the determination of the level of the offered quality.

(e) Cost of failure to provide offered quality

In the consideration of SLAs, revenue rebate may be made to the customer should one of the quality parameters be found not to meet the agreed level. Revenue considerations aside, the competitive threat arising from not meeting the offered level of quality will have detrimental effect on market share. The service providers credibility will be lost if continued nonconformance to offered quality occurs, with the resulting loss of market share.

(f) Population and usage criteria

The performance of a network service is often affected by the manner in which it is used. It is therefore necessary to understand the forecast population of users and how they use the services. For example, a usage profile which peaks during the day causes an overload which will severely degrade performance and hence the QoS. Also an unusual usage pattern which combines different services may cause undesirable feature interaction which degrades the performance of a number of services.

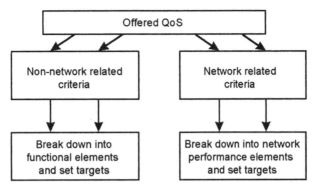

Figure 6.7   Offered-quality planning process

6.5.2 Specification of planning documents to meet offered quality

It was stated earlier in this chapter that the most extensive use of offered quality is in the derivation of planning documents and operational processes to meet the targeted level of service. The key activities in the translation of the offered quality into planning documents are illustrated in Figure 6.7.

The offered performance criteria are separated into network- and non-network-related criteria. Where some criteria are a mixture of both, the two elements have to be separated and included in the respective categories. The network-related criteria are broken down into element performance. The subject of specifying and managing network performance is treated in Chapters 10 and 11.

Non-network-related performance criteria are broken down to unique and manageable functional elements. The non-network-related parameters (from Table 6.1) for basic telephony are developed in Table 6.2.

Table 6.2   Non-network-related parameters for basic telephony (source: Table 6.1)

No.	Parameter Table 6.1	Parameter for offered quality
1	1	Target time to provide precontract information on services to customer, e.g. tariff, service availability, service features, choice of telephone features etc.
2	2	Target of a certain customer-satisfaction score for the service provider's professionalism
3	3	Target time for resolution of complaints, e.g. 90% of all complaints to be resolved in 4 hours
4	4	Target of, e.g., 90% of customers' faults reported to be put right in units of time (e.g. 4 hours)
5	5	Target of, e.g., 90% repairs right first time
6	8	Call-line-Identity (CLI) display options†

† *This parameter has both network- and non-network-related implications.*

The principal management issues related to the parameters in Table 6.2 are:

(i) *Time for precontract enquiries on the service*

Since basic telephony is a simple and ubiquitous service, customers will expect the service provider's point of contact to supply all the answers to their queries at once without any further internal consultation. The most common questions refer to availability of the service, connection charges, tariff charges, terms and conditions relating to the service and quality. However, for new more complex services, customers may be prepared for precontract enquiries to take several days. Large customers may even require service providers to tender competitively for provision of the service.

(ii) *Professionalism*

The service provider has to ensure that its customer-facing staff are trained to be courteous and business-like, sensitive to customer needs (e.g. special requirements such loudspeaking telephones for the hard of hearing). Additionally, customer-facing staff should have a basic understanding of the service provider's entire portfolio of products and services. The service provider should attempt to measure the customers' opinion of their professionalism on a regular basis. Any dissatisfaction on the part of customers could be put right by appropriate training.

(iii) *Complaint-resolution time*

Customers are usually sensitive to the time taken for the resolution of complaints. The time will usually vary from country to country due to environmental conditions and the maturity of the market. However, the optimum time for resolution should be obtained by customer survey, then benchmarked with the competition and finally compared with the current time taken. Remedial action is then taken if necessary.

(iv) *Time to repair*

Repeated surveys indicate that 'time to repair' is among customers' top ten quality requirements. The issues to be addressed by a service provider are current repair times within the business, world and principal competitors' benchmarks, customers' requirements, resources to offer the required time to repair and the penalties likely to be incurred should the repair times be not met.

(v) *Repairs not carried out right first time*

A recent survey carried out in the UK and Europe indicated that about 50% of business customers have experience of repairs not being carried out right first time. This is wasteful of both the customers' and the service provider's resources. The service provider is paying for the same problem to be rectified more than once and the customer suffers both financial loss and the inconvenience of repeat visits from the service provider. The only beneficiary could be the service providers' statistics if they are measuring 'total number of faults cleared'. The service provider will then be shielded from this quality requirement until the customer is lost. The issues to be addressed by the service provider are similar to those of 'time to repair' with the addition of ensuring that the repair staff are competently trained and managed effectively to achieve the desired results.

(vi) *Calling-line-identity requirements*

This parameter has both network and service implications. The service-related issue is providing operational resources within the service provider to ensure that records are correct and up to date.

More information is given in Chapter 17 on the practical management of the quality of a service.

6.6 Conclusions

The management of 'offered quality' by the service provider is one of the key elements of the successful overall management of quality. Offered quality forms the basis on which the service provider will be judged by its customers. To arrive at an offered quality, the service provider must be aware and have a full understanding not only of the customers' requirements but also of its own internal capability. The resulting service offer will be an agreed best fit between the customers' requirements and the service providers' capability. Once the offered quality levels have been established, the next step is to implement these and monitor the achieved performance; this is the topic of Chapter 7.

6.7 References

1   ZAIRI, M.: 'Competitive benchmarking' (Technical Communications Publishing Ltd., 1992)
2   CHERRETT P.: 'Practical Benchmarking', *Br. Telecomm. Eng.*, 1994, **12**, pp. 290–293
3   ITU-T Recommendation G.114: 'One way transmission time' (International Telecommuinications Union, 1996)

Chapter 7
Quality of service delivered

7.1 Introduction

Historically the *delivered* or *achieved* QoS has the oldest pedigree among the four viewpoints of quality. Of the four viewpoints, this is probably the best managed by the providers, albeit not as professionally as the customers would require. However, these were primarily technical performance, not quality parameters understandable by the majority of customers. Measurements on delivered performance of the more obvious technical parameters, such as transmission levels, noise levels in circuits, crosstalk etc. have been carried out on telecommunications networks for a long time. As progress was made in transmission capabilities, other parameters were added, such as delay and echo. As the quality of telephone instruments improved, sidetone was also added to the ever-growing list of quality parameters. Over the years, more and more parameter measures were added and a comprehensive set of performance parameters has evolved. These were based on objective measurements and were mainly associated with the service infrastructure (i.e. network technical performance). In due course other quality parameters were added, such as speed of provision, time to repair and similar operational parameters. Yet there are no internationally agreed definitions of performance parameters for this viewpoint. Today, there is a need to

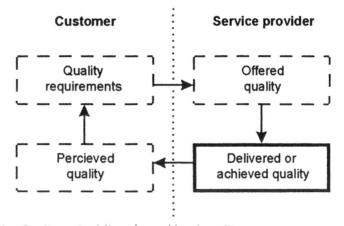

Figure 7.1   Quality cycle: delivered or achieved quality

understand clearly the meaning and significance of delivered quality and its relationship with the other quality viewpoints (i.e. the offered quality and the perception of customers).

In this chapter the characteristics of the delivered quality are stated, applications are identified and the principal management issues are examined.

7.2 Characteristics of delivered quality

The principal characteristics of the delivered quality are that:

- it is objective (i.e. it is a measure of the actual delivered performance from measurements, and therefore noncontentious);
- it may be expressed in either of two forms, as 'element performance' or as 'end-to-end performance'.

Delivered performance is usually estimated from measurements carried out on the network supporting the service and its relevant organisational support functions. Certain network-performance parameters may be monitored on a sampled basis, e.g. transmission performance. Others, such as exchange outage, are usually monitored continually to cover all nodes. Performance data on parameters which are non-network-related, e.g. time to repair, must be computed from 100% of incidents. If sampling is carried out on these parameters there is a risk of not providing customers with a true and fair view of achieved performance. The measurements are carried out by the network provider or the service provider. The parameters chosen for measurements must have a direct relationship to the parameters describing offered quality so that they can be compared.

The measured performance data may be presented in two ways:

(i) The delivered performance may be expressed on an element-by-element basis. This will be of particular benefit to the network provider and the service provider which can monitor the delivered performance against planned or specified element performance and take any necessary remedial action. Element performance is of little interest to customers.

(ii) The delivered performance may also be expressed on an end-to-end basis. This will be of particular use to customers and regulators, which may not be concerned with the element performance within networks. To the service provider it is an indication of how well or otherwise the service performance is delivered to the customer.

7.3 Applications of delivered quality data

The principal applications of delivered quality data are as follows:

(a) to monitor performance of network elements and operational capability of the service provider;

(b) to check the delivery of quality with agreed service level agreements (SLAs);

(c) to check whether regulatory requirements have been satisfied;

(d) to use as publicity material;

(e) to use as a reference point against which customers' perception of quality can be judged; and

(f) benchmarking.

(a) Monitoring performance of network elements and operational capability of the service provider

The data on delivered quality may be used by the network provider and service provider for variance analysis. In the absence of reasonable explanations for unaccounted variances in delivered performance from planned quality, it will be necessary to carry out investigative work. It is the service providers' responsibility to study the delivered performance of operational functions associated with a service. For example, if the end-to-end transmission quality of a national connection was designed to give a certain level of error-free performance and this level was not achieved, investigations may be initiated by the service provider through the network provider to identify the causes of this variation. Similarly, if the target 'time to repair' was 4 hours for 99% of faults on leased lines and only 78% of such faults were repaired on time, there is reason for investigation, this time by the service provider, not necessarily involving the network provider. Variance analysis for performance is based on the delivered performance and must be a standard activity for network and service providers.

(b) Check on compliance with SLAs

Delivered performance forms the basis for enforcing a SLA between the customer and the service provider. Audited performance data (see Section 7.4.4) may be used by the customers to pursue compensation claims in the cases of noncompliance to agreed performance levels within an SLA. The data may also be used by the provider to prove to customers that it is delivering to the targets set out in the SLA.

(c) Check against regulatory requirement

In certain countries, the regulator requires the service providers to publish data on delivered quality. In many such countries there is little or no punitive element for under achievement. The regulator is more concerned with publication of the data than the level of achievement for a particular service. The European Commission is still considering the introduction of common performance parameters to be published by the service providers of the European Union member countries. The regulator may consider introduction of punitive elements for under achievement

(d) Use as publicity material

Delivered performance data may be used as publicity material in the following areas:

- for customers nationally;
- for international comparisons of performance; and
- against competition.

Customers, naturally, want to know how well a service is performing. Such information may be published in the media. It may also be communicated individually by mailshots to organisations which have a special need for such information. Bodies such as the Organisation for Economic Co-operation and Development (OECD) and the European Telecommunications Network Operators (ETNO) regularly publish performance statistics (see Chapter 16). These comparisons could be of use to customers (mainly large companies) when choosing a provider for their communication services.

Perhaps, the single most useful benefit to customers is to identify providers which offer the best quality. Professionally presented quality data will be welcomed by the customers. Those service providers which are reluctant to publish data on quality may be seen by the market as trying to hide poor quality.

(e) Use as reference point against customer perception of quality

The delivered quality forms the basis for the service provider to evaluate the customers' perception of quality. Since delivered quality is noncontentious, if monitored professionally and with an audit trail, unexpected customer-perceived quality ratings can be investigated. The evaluation of customer perception and the relationship to delivered quality are dealt with in Chapter 8.

(f) Benchmarking

When a service provider wishes to ascertain the world's best (benchmark) for performance, delivered quality data are the most often sought. These data are compared with the service provider's own performance to determine the gap that may need to be closed. Benchmarking is dealt in more detail in Section 6.5.1.

7.4 Management of delivered quality

7.4.1 General

The responsibility for the management of delivered quality lies with the service provider. It is the service providers' responsibility to liaise with the network providers for network-related performance.

7.4.2 Specification of monitoring systems

The process of specification of monitoring systems is illustrated in Figure 7.2 Specification of monitoring systems involves the following principal steps:

(a) determination of parameters to be monitored; and
(b) specification of monitoring systems for each parameter.

The principal activities in the management of delivered quality are:

- specification of monitoring systems;
- determination of end-to-end performance from measurements;
- audit of delivered quality data;
- review of delivered performance; and
- roles and relationships between service providers and network providers.

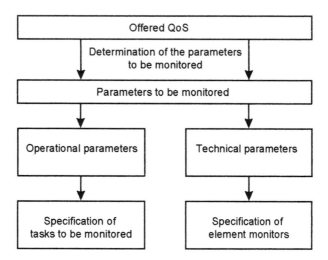

Figure 7.2   Process for the specification of monitoring systems

(a) Determination of parameters to be monitored

Determination of parameters to be monitored depends on whether the parameters are primarily for use outside or inside the service provider's organisation. Choice of performance parameters for external use would further be influenced by:

- customer preferred parameters on which delivered quality data are to be published regularly;
- requirements of regulators; and
- service providers' obligations towards organisations for international comparisons of performance.

Customers' preferred QoS parameters for regular publication are best captured during the determination of their requirements on quality (see Section 4.5). A typical set of such parameters is illustrated in Table 7.1

Table 7.1 Typical set of performance parameters on which customers' require delivered results to be reported on a regular basis

Number	Parameter
1	Time to repair
2	Billing accuracy
3	Speed of provision
4	Fault rates
5	Service availability
6	Complaint-resolution time
7	Adherence to CLI requirements
8	Number of complaints
9	Connection quality
10	Professionalism

A number of parameters on which customers require delivered performance to be published may be obtained on a service-by-service basis. Some of these parameters may be combined, e.g. 'repair time' and 'resolution of complaints', even though the publication of statistical information may require separation of data on a service-by-service basis. In addition, individual customer requirements may have to be accommodated. This is particularly true for SLAs. Business customers which insist on SLAs will state parameters on which delivered performance is to be reported. The service provider will be faced with an array of quality parameters to cover the complete customer base. However, through careful combination of activities it should be possible to achieve economies of scale in allocating resources to establish the relevant systems for monitoring the parameters.

The requirements of regulators would in many cases reflect the customers' requirements. However, the regulator's requirements often lag many of the leading businesses' requirements and the service provider could be faced with a regulatory demand for publication of performance parameters which could be different from the requirements of its customers.

A third consideration is service providers' commitment or obligation to provide data for organisations such as OECD and ETNO which attempt to publish comparative performance data from their member countries. Neither of these bodies has yet succeeded in obtaining performance statistics from all its member countries on agreed definitions of performance parameters but most service providers have supplied information matching closely to the suggested definitions.

(b) Monitoring systems

After the determination of performance parameters on which delivered performance should be published, they must be segregated under non-network-related and network-related categories. Certain parameters may contain both operational and network components.

Monitoring systems for non-network related parameters
From Table 7.1 the non-network-related parameters are:

- time to repair*
- billing accuracy*
- speed of provision*
- fault rates*
- complaints resolution time
- adherence to calling line identity (CLI) requirements*
- number of complaints
- professionalism

Asterisked parameters have network implications.

The guiding principles in establishing monitoring systems for operational parameters are relatively straightforward. For example, to produce statistics related to 'time to repair' the following raw data are needed:

- record of time at the instant a fault was reported;
- record of time at the instant the fault was cleared, i.e. when the customer was notified of the capability to use the service again;
- geographic identification of the fault incidence.

From these records a number of statistics can be determined. Examples are:

- mean time to repair;
- percentage of repairs carried out within various times
  (e.g. 5 hours, 4 hours, 3 hours, 2 hours, 1 hour);
- standard deviation of the distribution of repair times;
- performance on a national basis and for selected geographical areas;
- 'hot' spots (areas where long times for repair exist).

Production of this statistical information need not be expensive. The basic data may be submitted to a computer program to produce the above statistical information.

The guiding principles for monitoring systems for nonquantitative parameters (e.g. professionalism) are more tenuous. The obvious method of expressing the professionalism of a service provider is to quote a customer-satisfaction rating for this particular parameter (customer perception is dealt with in Chapter 8). Customer ratings may also be used for other nonquantitative parameters such as flexibility, courteousness and politeness of staff, sensitivity to customers' needs etc. on the part of service providers. However, wherever possible, quantitative measures should be used and backed up by customer ratings.

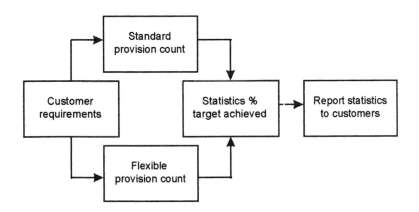

Figure 7.3   Measure of provision flexibility

If a provider wished to measure flexibility of provision of a service, this could be delivered by the introduction of a simple counting system as shown in Figure 7.3.

When a customer requires provision of a service, there are two possible counts which may take place, one against the standard provision date quoted by the provider, and the second against a flexible-provision date. Both are monitored against achievement and statistics produced. For example, if there were 100 requests for service to be provided outside the standard provision time, these would be counted separately. If 79 of these provision requests were completed by the customers' required date and the other 21 were not, the flexible provision performance would be 79%. This could then be compared with the target and the resulting information supplied to the customer as proof of service quality.

Specification of monitoring systems for network related parameters
Network-related parameters are described in Chapter 10. When specifying monitoring systems, the following criteria need to be addressed:

- identification of the unique elements whose performances are to be monitored;
- estimation of the sample sizes and frequency of measurement; and
- method of measurement.

Consider the performance parameter availability. Figure 7.4 illustrates conceptually an end-to-end connection. Links E_3, E_5 and E_7 may be traffic routes, which are usually monitored against call congestion and outage, or transmission routes which are monitored for selected transmission technical performance, e.g. error and outage, or signalling routes which are monitored for signalling overload and outage. E_1 and E_9 are local access lines (or local loops) and it is impractical for dedicated monitoring systems to be employed due to the potentially high cost and the relatively low incidence of faults per

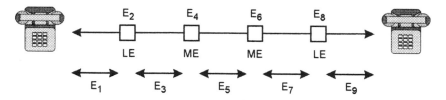

Figure 7.4   Elements of a hypothetical connection

E_1 = local line from customer to local telephone exchange
E_2 = local telephone exchange
E_3 = link between local telephone exchange and main telephone exchange
E_4 = main telephone exchange
E_5 = link between main telephone exchanges
E_6 = main telephone exchange
E_7 = link between main telephone exchange and local telephone exchange
E_8 = local telephone exchange
E_9 = local line between the local telephone exchange and the customer

line. However, in special circumstances monitoring equipment could be installed in the local access network and outages monitored. In most cases, only reported outages are used in the calculation of end-to-end performance. Links E_3, E_5 and E_7 carry a high volume of traffic and are not confined to providing service to one customer. Hence, these must be monitored for outage. Nodes E_2, E_4, E_6 and E_8 need to be monitored regularly for outages. Some exchanges have their own internal monitors to indicate the outage durations. Where this is not built-in, it is necessary to provide an external device, to be triggered by the outage and reset by service resumption to note the duration and date of the outage. All nodes are monitored for traffic congestion and when thresholds are exceeded they trigger management action.

7.4.3 Determination of end-to-end performance from observations

In the publication of delivered quality for the benefit of customers and regulators, the performance statistics of certain parameters are best expressed on an end-to-end basis. The estimation of end-to-end performance of operational or non-network-related parameters is straightforward. However, the computation of technical or network-related parameters is more involved. Availability of service could be affected by a fault occuring on a number of different elements within the network. The fault history of an element used in the end-to-end-performance calculation is of vital interest to the provider but of no concern to the customer. As far as the customer is concerned, it is the availability of the service that matters, and not any individual element's performance used in the calculation of end-to-end performance. The principle of such a calculation is illustrated with the connection set up of Figure 7.4.

If the outages in hours for the different elements are represented by O_1–O_9 for the elements E_1–E_9 respectively, the total end to end outage may be estimated as total outage O_T $\qquad = O_1 + O_2 + \ldots\ldots + O_9$

$$= \sum_{n=1}^{n=9} O_n$$

percentage availability $= \left(1 - \dfrac{O_T}{365 \times 24} \right) \times 100$ over a period of one year

This estimation makes the following assumptions:

- the outages of the individual elements are mutually exclusive;
- all traffic between E_1 and E_9 is carried on this route.

However, in practice, the solution is complicated by the following:

- outages may not be mutually exclusive;
- local-telephone-exchange and main-telephone-exchange outages may not affect all traffic that pass through these exchanges;
- outages may be automatically restored before they affect customers, e.g. transmission service protection and traffic management (e.g. rerouting).

With connection quality, the estimation is more complex. The degrading parameters of a connection are: transmission loss, stability, echo, delay, error, slip, jitter, wander, sidetone, noise, quantisation distortion, loss distortion with frequency, group delay distortion and crosstalk. The implications and management of these parameters are described in Chapter 10. The appropriate parameters for the type of connection need to be identified (e.g. on an analogue connection, error rates, slip, jitter, wander and quantisation distortion will not be applicable). The cumulative effect of degrading parameters has been the subject of many studies, as these parameters do not combine linearly. Recommendations in the P11 series illustrate the effect of various parameters on the transmission quality to the customer. The combined effect of other parameters contributing to overall transmission quality are either under study or studies are yet to be carried out.

In the estimation of end-to-end technical performance parameters, based on a sample of circuits measured, the need for confidence levels and the maximum systematic errors should be evaluated. The service provider must judge whether additional data would be of benefit to customers. It will certainly be of interest to the service provider and to students and professionals in the telecommunications industry.

Many measurements are subject to error, and the final result is expressed as being correct to within the 'maximum systematic error'. The contributory factors to the systematic errors are errors or tolerances of the measuring devices, uncertainty errors and errors due to changes in environmental conditions. If the total error is the sum of two or more components, the total can be expressed as follows.

If the nominal value is expressed as $y = u + v$ and if the measurement errors are du and dv, respectively, the corresponding maximum error in y is $\pm$ (du + dv). If the nominal value is expressed as $y = u \times v$ and the relative errors are du and dv, respectively, then the maximum error dy/y is expressed as $\pm$ $(du/u)+(dv/v)$. If the nominal value is expressed as $y = u/v$ and the errors are du and dv, the maximum systematic error is also given by $\pm$ $(du/u)+(dv/v)$. For a treatment of errors due to measurement, see Topping [2] or Buckingham and Price [3].

A service provider or network provider has to address the issues of setting up measurement systems to compute delivered network performance. Rogers and Hand describe how some of such issues have been addressed in British Telecommunications plc. [4]. Measurement of network performance needs to be refined to meet customers' end-to-end performance requirements.

Processing of delivered performance for publication
Relevant supporting information must be provided when end-to-end performance is published for external use. Such supporting information could include any abnormal performance which may have occurred during the period of the report, and their causes, For example major exchange outages caused by localised fire, flood, earthquake etc. and serious weather-related operational difficulties which contribute to poor service availability may be stated. If the availability performance is averaged out for the whole country, the localised effect would be masked. If a detailed explanation is given by the service provider and the national figures with and without these outages are stated, these could be more meaningful to the customers.

7.4.4 Audit of delivered performance data

The delivered performance data, when computed for internal use within the service provider's or the network provider's organisation need not be audited externally. However, if these data are to be used for customers' benefit or by any other body external to the supplier, it adds confidence to these recipients if the data are audited by an independent body. When punitive measures are to be authorised for under achievement, it is desirable for performance figures to be audited.

Guidelines for auditing have been established by the international standardising bodies. *ISO 10011* describes these [5]. Three stages of auditing are normally required. These are:

- systems audit;
- compliance audit;
- follow-up audit.

In the systems audit, examination is carried out for the presence of detailed documentation and its conformance to the requirements for the measurement of the parameter components. This review must include a verification of the criteria for capturing the events associated with that particular parameter.

The compliance audit is a test of implementation of the system and examination of its efficiency and effectiveness. This audit consists of on-site verification of objective evidence by the independent body. The follow-up audit is a review of deficiencies which have been identified at the audit stage. This audit determines whether or not the corrective actions agreed on have been completed. The frequency of audit visits will depend on how well the system has been documented and put into practice. Good working systems should not require more than one visit a year.

7.4.5 Review of delivered performance

The delivered-performance statistics ought to be reviewed if any of the following implications are pertinent:

- assessment of delivered performance shows a variance from the planned levels of quality; upon analysis, there could be two outcomes: one is that planned levels are inadequate and the other is that the measurement methods are suspect.
- comparisons against customers' perception of quality show there is a substantial lack of correlation between customers' and delivered qualities.

The delivered performance is normally monitored by the network or service provider. The data are compared with the planned levels of quality. Where there is significant variance between delivered and planned quality, investigations must be initiated to identify the causes of these variances. Examples of possible causes are:

(a) faulty measuring equipment;
(b) adverse conditions affecting performance; and
(c) optimistic planning assumptions.

It is important to identify the causes of the variance in order to carry out remedial actions and to restore the delivered performance to its planned levels whenever possible. However, certain measurements for sources using leading-edge technology may prove either inadequate or unsatisfactory and may need to be replaced in the light of experience.

In Chapter 8, customer perception of quality is compared with the delivered QoS and the implications for the provider.

7.4.6 Service- and network-providers' roles and responsibilities

Where service providers are not network providers, it is necessary for the relationship between these bodies to be discussed and agreed amongst themselves. The service provider is expected to deal with all aspects of quality with the customer. In addition, the service provider has the responsibility to ensure that the network provider agrees and monitors specific performance parameters within the network to ensure that the correct end-to-end service quality is maintained. The range of parameters to be monitored and the elements to be measured must be agreed between the two parties for effective management of the delivered performance. Such agreement of mutual responsibilities should also be extended to international network and service providers when services are provided beyond the border of a country.

7.5 Conclusions

Management of delivered quality, even though the most prevalent form of quality management among the service providers, requires better professionalism for optimum effectiveness. Scientific determination of relevant sample sizes, appropriate parameters (for optimum benefit to customers, regulators and the service providers) and specification of the most efficient monitoring systems are parts of this process. Determination of the delivered performance leads to the next steps in the management of quality: the management of customers' perception of quality, determination of the variances and establishment of remedial actions. These are dealt with in Chapter 8.

7.6 References

1   ITU-T Rec. P11: 'Effect of transmission impairments' (International Telecommunications Union, 1993)
2   TOPPING, T.: 'Errors of observation and their treatment'. The Institute of Physics and the Physical Society monographs, (Chapman and Hall, 1966)
3   BUCKINGHAM, H., and Price, E. M.: 'Principles of electrical measurements' (English Universities Press, 1966)
4   ROGERS, D., and HAND, D.: 'Network measurement and performance', *Br. Telecomm. Eng.*, 1995, **14**, pp. 5-11
5   ISO 10011: 'Guidelines for auditing quality systems. Part 1: Auditing; Part 2: Qualification criteria for quality system auditors; Part 3: Management of audit programs' (International Standards Organisation, 1993)

Chapter 8

Customers' perception of quality of service

8.1 Introduction

The starting point for this part of the quality cycle should be, 'customer perception is reality', even if that perception does not always mirror reality [1]. It is important for a service provider to be aware of the implications for revenue, profitability and market share of adverse customer perception ratings.

A customer's perception of quality is a judgement made by the customer about the overall excellence, or otherwise, of the delivered product or service. Therefore, customer perception is an essential factor in the successful management of a product or service. Customer perception is the main criterion by which service providers can assess and measure the true value of the quality they deliver. Established industries, particularly in competitive environments, regularly carry out customer-perception surveys to assess how well (or otherwise) their products and services are being perceived by

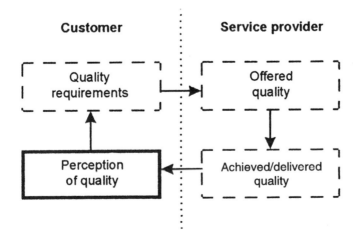

Figure 8.1   The quality cycle — customer perception

customers. Indeed, the service provider's decision-making process would be incomplete without this information. However, the information contained within customer-perceptions surveys is usually very subjective and therefore likely to change over time with changes in the industry structure, environment, technology, competitive activity and good or bad media commentary. Therefore, a service provider must constantly review customer perceptions of its products and services and take any remedial action necessary to improve or correct the current perceptions.

Customer-perception surveys usually fall into two parts. One asks the customer questions built around 'are you happy with the product or service?' and the other is concerned with areas of dissatisfaction. There is a great many statistical data quoted to describe customer dissatisfaction with quality and their behaviour towards a service provider. For example, according to one study, as detailed by Brown [2]:

> '70% of customers dissatisfied with a service will go elsewhere, but only 5% will tell you they are unhappy ... Dissatisfied customers tell an average of 10 people about their poor experiences while satisfied customers will tell only five. It costs up to five times as much money to attract a new customer as to keep an existing one, but 95% of dissatisfied customers will stay loyal if they are handled properly.'

These figures will normally vary with the type of service, from country to country, with competition, circumstances and time.

Irrespective of the precise magnitude of such figures, it is clear that it is very costly to have dissatisfied customers. For this reason, customer perception of quality forms an important part of the overall management of the quality of a product or service. This argument also applies to telecommunications. The principal objective in the successful management of customers' perception is to assess their subjective opinion in terms of quantitative ratings, identify negative perception and attempt to rectify this to the mutual benefit of customer and the service provider.

8.2 Characteristics of customer perception

8.2.1 General

The principal characteristics of customers' perceived quality are:

- it has varying degrees of subjectivity;
- it is expressed as opinion ratings quantitatively.

8.2.2 Subjective elements

Customers' perceptions could be influenced by a variety of factors which will affect their judgement of quality. Examples of these influencing factors are:

(a) customers' 'awareness' of telecommunications services;

(b) customers' expectations;
(c) experienced quality;
(d) customers' recent experiences;
(e) advertising;
(f) nature of the opinion survey.

(a) Customers' awareness

There are many elements in the complex array of influences which have led to increased customer awareness.

Business customers have well developed, clearly defined needs and their choice of service provider will be principally governed by their level of confidence in an operator to satisfy them and hence support their business activities.

Conversely, personal customers may be less discerning and accept more basic, reliable communications services. They are more influenced by softer 'service-surround' issues such as courteous customer service and an efficient, effective repair service. Price is an issue for them but few personal customers would go to the inconvenience of switching operators on the basis of a nominal price differential alone. Any such move would be either opportunistic, such as during a house move, or be prompted by an emotional reaction to a negative, unsatisfactory experience with the service provider.

Customer perception may also vary depending on customer type (i.e. social group), customer segment and sector and geographical location. For example, Figures 8.2 and 8.3 indicate the difference in customer perception

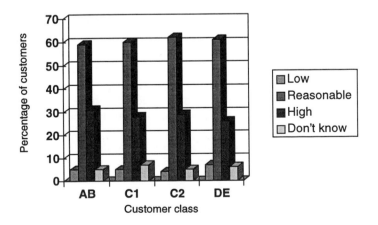

Figure 8.2   Graph of customer perception of local call prices by social group (source: internal study)

of local call charges within the UK, based upon a rating of low, reasonable, high and don't know.

Figure 8.2 shows that there is no correlation between the perception of cost and disposable income, with the possible exception of people at the lowest income levels. It is perhaps arguable that adverse opinions of the highest earners are caused by a greater awareness of competitor offerings gained through experience at work.

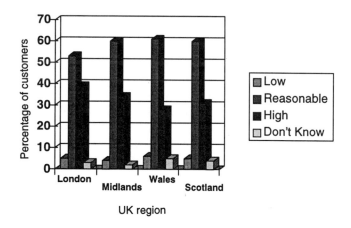

Figure 8.3   Graph of customer perception of local call prices by region (source: internal study)

The results for perception of local calls are influenced by customer beliefs about other telecommunications operators. Also, UK competitors' price communications have an impact, as does the belief that operators outside the UK offer free local calls.

Figure 8.4 shows some of the influences which combine to shape customer awareness. To consider briefly the major areas of influence:

Evidence: The customer is confronted with many forms of evidence, from external publications and media, to publicity, advertising and management reports produced by the service providers themselves.

Badly presented bills or management reports awash with meaningless information will evoke a negative reaction from the customer. By paying attention to those aspects of the customer experience concerned with the generation and consumption of 'evidence', the service provider has a powerful tool through which to influence, win and retain customers in a competitive environment.

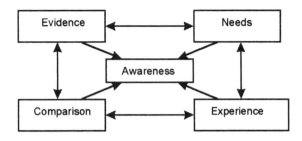

Figure 8.4   Influences on customer awareness (source: adapted from Reference 3)

Comparison: The criteria by which one operator is compared with another will vary over time and with the needs of the customer in question. As competition increases, so customers' comparisons will have a greater effect on their perception. The two most likely areas customers will attempt to compare are price and quality.

Needs: The specific needs of a customer will have an impact on perception. A personal customer may need to receive separate bills for a number of individuals (i.e. teenage children) who all make calls from a single telephone line. During certain periods in their life, such as family illness, it may be important for them to have 100% service availability. Hence their needs will vary with their circumstances.

A business customer may need a structured, customised communications plan, requiring services and service combinations outside the service provider's portfolio or technical capability. They may need a particular style, frequency or delivery of billing information to assist in dynamic business accounting.

Experience: The most direct, and therefore the most potent, form of influence on customer perception is personal experience. Research has shown that customers' mood, involvement level and the quality of the experience with a product service or service provider have a significant effect on their future buying intentions. Customers in good moods tend to evaluate good experiences still better, and customers in a bad mood tend to exaggerate bad experiences and make them even worse.

Each customer will have certain expectations against which they will be most sensitive to the quality of their experience with a product or service. If the service provider is not adequately attuned to those expectations, the quality of the experience for the customer will be low.

(b) Customer expectations

In a study by Zeithaml, Parasuraman and Berry [4], it is shown that perceived quality is related to expected quality, as shown in Figure 8.5. The model defines the key determinants of expected quality as 'word-of-mouth communications', 'personal needs', 'past experience' and 'external communications' from the service provider.

Word-of-mouth: This covers what customers hear from other customers, i.e. recommendations, which may be good or bad and are directly outside the control of the service provider.

Personal needs: This embraces a multitude of criteria based on a customer's personal characteristics and circumstances which are not totally under the control of the service provider. A service provider will not be able to meet the totality of a customer's needs (e.g. total customisation) economically.

Past experience: This will encapsulate an individual customer's entire past experience of a particular service provider. Whether this is a good or bad experience will be a matter of fact and the starting point for all future dealings. Therefore, this determinant will be outside the direct control of the service provider.

External communications: This covers all the contact between the service provider and the customer, including advertising, media coverage, promotional material, sales and enquiry contact. All of these except media coverage are under the direct control of the service provider.

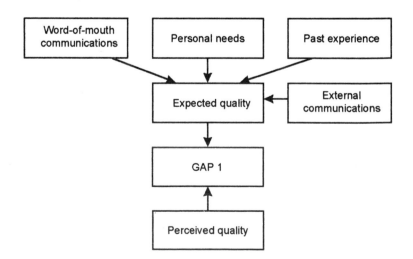

Figure 8.5   Customers' expected and perceived quality gap (source: adapted from [4])

The model in Figure 8.5 shows the gap between expected and perceived quality (gap 1). It follows that, by good management of the key determents by the service provider, this gap can be reduced. However, the amount by which the gap can be reduced will depend on the service provider's control over the determinants. For example, external communications can be made up of advertising (directly under the service provider's control) and media coverage (not directly controlled by the service provider).

According to Gronroos [5], good perceived quality is obtained when the quality experienced is the same as the quality expected, as illustrated in Figure 8.6. From this model it can be seen that the expected quality is linked to the traditional marketing activities undertaken by a service provider. Therefore, to some extent the service provider can set the expectations of its customers and this will then have an effect on the perceived quality. However, image, word-of-mouth and customer needs are not entirely under the control of the service provider, as explained above.

The experienced quality is affected by a number of determinants as shown in Figure 8.6:

Achieved quality: Achieved quality is made up of technical performance and service quality. Technical performance is a technical measure of how a product or service performs during its life cycle (e.g. does it meet all its technical performance parameters and targets?). The magnitude of this determinant will depend on an individual customer's bias towards the technical aspects of the product or service. Therefore, the influence of this determinant on experienced quality may vary from customer to customer.

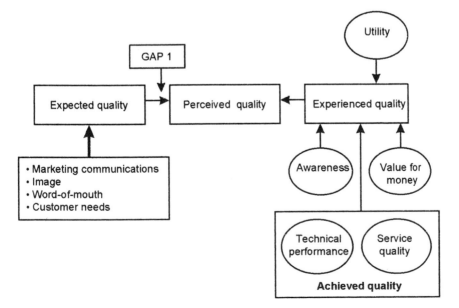

Figure 8.6   Perceived-quality model (source: adapted from Reference 5)

Service quality is the aspect of experienced quality which reflects how well the product or service is supported, and is explained in detail in Chapter 9. It must also be stressed that the aspects of service quality which are important to customers and affect this determinant will again vary from customer to customer.

Awareness: Experienced quality will also be influenced by a customer's awareness of similar products and services on the market. However, the affect of this on experienced quality will be influenced by the factors forming 'expected quality' explained in relation to Figure 8.5.

Value for money: This is a subjective judgement made by the customer when considering all the aspects of quality including price. If a customer feels that he or she has, or has not, experienced value for money, then it also follows that subjectively the experienced quality may be affected.

Utility: This refers to the 'usefulness' of the product or service (e.g. does it do all the things it is supposed to do?). The limits of this determinant can be set by the service provider, with careful consideration of their external communications.

It can be seen from Gronroos' model [5] and the adaptation in Figure 8.6 that perceived quality is a function of expected quality and experienced quality. If customers' expectations are unrealistic, the total perceived quality will be low, even if the experienced quality is good. This shows that it is important to understand the linkages and the determinants which make up both 'expected' and 'experienced' quality. The model serves as a useful tool to focus a service provider's attention on the key determinants which need to be carefully managed in order to improve customers' perceptions of their products and services.

(c) Experienced quality

In an ideal world, the perceived quality should correspond directly to the quality experienced (which includes achieved quality) by the customer. Even though customers may require improvement in quality, their perception is more likely to correlate with the quality actually experienced, especially if they have a service level agreement (SLA) with the service provider. Customers are more likely to monitor the experienced quality when there is an SLA with the service provider. Their expectations of quality will reflect the contents of that SLA.

(d) Recent experiences

A customer's perception will be influenced by his or her most recent experience with the service provider, the timing of that experience and its perceived value to the customer (i.e. whether it was a good or bad experience).

A recent bad experience is very likely to degrade a customer's opinion, sometimes unduly. Similarly, a good experience may negate recent bad experiences, depending on the quality of the good experience. For a bad experience to be forgotten, a number of counteractive good experiences may be required. In telephone interviews and personal interviews, it is easier to assess the depth of these adverse experiences and assess how much these have influenced the opinion of a customer. Such an assessment would enable a more realistic correlation to be made between the perceived and experienced quality.

(e) Advertising

It has been found that advertising has an impact on customer perception. Advertising is an excellent tool with which to influence perception, because the messages are directly under the control of the service provider. Image and 'word of mouth' are only indirectly controlled by the service provider, because they are based on the past performance of a product, service or service provider experienced by either the customer or someone close to the customer. Research in the retail industry has shown that comparative advertising (advertising which compares product performance) and price have an effect on perceived quality. In the work carried out by Gotlieb and Sarel [6], three hypotheses were put forward with supporting evidence. These were:

(i) A direct-comparative advertisement will have a more positive impact on perceived quality of a new brand than will a noncomparative advertisement.
(ii) There will be an interaction effect of source credibility and price on the perceived quality of a new brand (the source here is the person who delivers the message in the advertisement, the person's credibility being the influential factor).
(iii) There will be a three-way interaction effecting the type of advertisement (i.e. direct-comparative against noncomparative), source credibility and price on purchase intention toward a new brand.

A direct-comparative advertisement will have a more positive impact on perceived quality of a new brand than will a noncomparative advertisement. This impact is particularly great for a new brand and when the comparison is made against a market leader. When the source is highly credible, the consumers are less likely to discount the message of the advertisement. Consequently the influence of price is moderated and they are more likely to use multiple cues (e.g. product attributes, price and source credibility) to judge quality of a new brand. Conversely, when the source has low credibility, the influence of price on perceived quality is less likely to be moderated by other cues in the advertisement and consumers are more likely to discount the message.

The applicability of the above hypotheses to telecommunications services is unknown. However, there is reason to believe that the first two hypotheses would be relevant in a customer's perception of quality of a telecommunications service.

(f) Nature of survey

Customer surveys are subjective by nature and the results should therefore be interpreted with caution. The three most popular ways of conducting customer perception surveys are postal questionnaire, telephone interview and person-to-person interview. The wording of a questionnaire, the way questions are asked and the mood of the interviewer and interviewee will have a significant impact on the answers given and the ratings associated with those answers. A customer will also be likely to give slightly different opinion scores depending on the viewpoint of the interviewer. If the interviewer is the supplier, user group, regulator or an independent consultant, the interviewee's opinions and corresponding opinion ratings could be significantly different.

8.2.3 Customer-opinion ratings

Customers' perception of quality is expressed qualitatively (e.g. I think the service is good), but to turn this type of statement into meaningful data covering a large number of customers this perception must be expressed quantitatively. For example, Figure 8.7 shows the concept of illustrating qualitative statements with quantitative values suitable for further analysis. The rating attached to statements such as 'good' enables the service provider to perform mathematical computations on the qualitative data, for further analysis.

Voice quality

The service provider will attempt to provide 100% of connections of sufficient quality to enable you and the person at the other end of the connection to understand each other without difficulty. However, this may not always be possible. How would you rate the quality of your voice service?

Excellent	Very good	Good	Average	Fair	Poor
6	5	4	3	2	1

On a rating of 1-6 please indicate how important this parameter is to you (ignore the present performance you receive).

Very important					Unimportant
6	5	4	3	2	1

Tick appropriate box

Figure 8.7   Method to quantify qualitative opinion

8.3 Applications

The principal applications of customer-perception ratings on quality are:

- as an indicator of market perception (i.e. 70% of customers find 'call quality' good);
- to compare service providers' quality (i.e. 70% of customers rate service provider A's 'call quality' as very good while only 30% of customers rate B's 'call quality' as good);
- to establish the gap between expected and perceived quality (i.e. 90% of customers expect 'call quality' to be very good while only 30% of customers perceive 'call quality' as being very good);
- to prioritise quality improvements for maximum benefit to both customers and the service provider (i.e. 90% of customers expect 'call quality' to be very good while only 30% of customers perceive 'call quality' as very good, but 80% of customers rate the priority of 'call quality' as very high).

By the successful application of perceived-quality data and quality programmes coupled with good advertising, a service provider could considerably increase customer perception of its products and services. It can then enhance company image with only a marginal increase in cost, because it can concentrate on parameters with a very high customer priority.

8.4 Management of customers' perception of quality

8.4.1 General

The key issues to be addressed in the management of perceived quality are:

- the measurement of customer perception;
- perceived-quality variance analysis; and
- modification of quality programmes.

8.4.2 Measurement of customer perception

The principal activities in the measurement of the customer perception may be grouped as:

(a) selection of service or group of services to be surveyed;
(b) choice of performance criteria to be surveyed;
(c) survey samples and size;
(d) questionnaire design and implementation;
(e) publication of results.

(a) Selection of service or group of services

Surveys may be carried out to achieve various objectives, for example to gain an overall opinion of a service provider. This need not involve detailed questions on service performance but could concentrate on general areas and ask whether respondents are satisfied with the overall performance of the service provider. Such surveys are of limited value and concentrate on company image and brand value rather than the performance of any specific product or service.

Specific product and service surveys are at a more detailed level and it is usually necessary to survey specific customer groupings (e.g. market segments and sectors). Such market segments and sectors have unique requirements as well as specific opinions of a service provider, probably based on previous perceptions of products and services and also based on perceptions of the competition's products and services. The perception of the competitor's quality may also be based purely on a perception of that company, as the customer may not have had previous experience with that supplier's products or services. These surveys are designed to focus on specific areas of quality on a specific product or service, and hence may have only a minimal effect on the overall perception of the service provider.

The service provider may carry out customer surveys to study the customer's views of their competitors and use this information as a marketing tool. The choice of customer groups to be targeted would depend on the service or group of services being compared.

A clearly-defined survey objective will focus the issues and assist the planning of the survey for maximum mutual benefit of the customers and service provider.

(b) Choice of performance criteria

The choice of performance criteria on which customer perception is to be measured will depend on the purpose of the survey. If a general opinion of the service provider is the objective, the survey questions will concentrate on the image of the company. In this case a method proposed by Zeithaml *et al.* [4] could be used. The method identifies 22 performance criteria which are of concern to customers. These have been grouped under five headings: reliability, responsiveness, tangibles, assurance and empathy, as shown in Table 8.1.

They recommend that survey questions are asked in two stages. The first is to ascertain the customer's *importance* rating for each of the 22 criteria to identify their expectations. In the second, the same questions are asked slightly reworded to seek respondents' ratings for perceived performance. Before the resulting data are analysed, customers are asked to distribute 100 points to five questions, representing the five groupings shown in Table 8.1. The relative weights of the five groupings are established from the allocated points. The interview data are then prioritised for all parameters, together with gaps between expected and perceived quality. Such prioritisation, it is claimed, will enable the service provider to allocate resources in areas where they are most needed to improve quality.

The approach by Zeithaml *et al.*, though very methodical, should be adapted for the telecommunications industry and also to suit the relevant country culture.

Table 8.1   SERVQUAL questions and groupings [4]

Groupings	Expectation score from 1 to 7	Perception score from 1 to 7
Tangibles (SERVQUAL questions 1–4)	Example: Excellent .... companies will have modern-looking equipment.	Example: XYZ Co. has modern-looking equipment.
Reliability (SERVQUAL questions 5–9)	Example: When excellent .... companies promise to do something by a certain time, they will do so.	Example: When XYZ Co. promises to do something by a certain time, they do so.
Responsiveness (SERVQUAL questions 10–13)	Example: Employees in excellent ... companies will never be too busy to respond to customers' requests.	Example: Employees in XYZ Co. are never too busy to respond to your requests.
Assurance (SERVQUAL questions 14-17)	Example: Employees in excellent .... companies will have the knowledge to answer customers' questions.	Example: Employees in XYZ Co. have the knowledge to answer your questions.
Empathy (SERVQUAL questions 18–22)	Example: The employees of excellent .... companies will understand the specific needs of their customers.	Example: Employees in XYZ Co. understand your specific needs.

If the purpose is to improve the performance product or service, then the criteria or parameters chosen should also be of concern to the customers. The parameters indicated by the customers during the capture stage (Chapter 5) may be considered most pertinent for the determination of customer perception.

(c) Survey samples and samples size

To identify the customers (samples) and the number of such customers (sample size) the following criteria must be considered:

- customers which have had recent quality-related experience with the supplier;
- customers which have had recent contact with the supplier; and
- customers chosen purely on a random basis (from the customer grouping identified for the survey, e.g. transport sector).

Customers which have had recent quality-related experience in the area to be studied with the supplier (irrespective of whether the experience was good or bad) are best suited for inclusion within the sample. The available sample size of such customers may be small but this would be easily be offset by the freshness of their experience.

Secondly, customers could be chosen which have had recent contact with the service provider, though not necessarily related to quality. Care must be taken in choosing these customers as they will not want to be continually researched, even if they perceive benefit from the research. The usefulness of the responses given by this sample may be of limited value due to the relevance of their recent experience. This method of selection is the next best and should only be chosen after the first category of customers has been exhausted and the sample size is not considered sufficient.

The third category of selection is from the total customer population of a particular customer grouping (i.e. all customers within the retail sector). This group will include customers from the above two categories and also those which have not had any experience with the supplier, and will constitute by far the largest sample size. Choice of samples from this group may also be of limited value as their perception will be based on company image and not past experience and this could not be representative if the results are applied to particular product or service-quality enhancement.

The customer-sample size for any survey will depend on the customer grouping chosen. A sample size of approximately 30–50 would be sufficient to provide good results if all the customers within the sample have had a recent quality experience in the area to be studied. However, as we move to the other two categories of customer groupings, the available sample size grows. If the range of responses varies significantly the sample size will need to be increased to provide statistically robust results. Customer responses to opinion rating should be studied from both a statistical viewpoint and a psychological viewpoint.

(d) Questionnaire design and implementation

The questionnaire should contain questions which reflect the objectives of the survey, as discussed in Section 8.4.2(a). The principal guidelines indicated in Section 5.4.2 for the capture of customers' quality requirements may be applied to the determination of customers' perception of quality. The question design and structure are critical to the successful management of this viewpoint of quality. Badly designed questionnaires will only reveal part of the 'whole truth' and could show the service provider to be either disproportionally good or bad. If, for example, the structure and content of the questionnaire tended to hide or diminish customer dissatisfaction with a product or service, the service provider would be lulled into a false sense of security with the belief that it was providing better quality than it in fact was. The worst realisation of this fact comes when the customer takes its custom elsewhere, with the consequential loss of revenue. The converse of this is when the structure and content of the questionnaire produce false results which show the product, service or service provider as underperforming in terms of customer satisfaction. The service provider will then tend to waste valuable resource in attempting to fix a problem which does not exist. Either of these two scenarios will have the effect of reducing the service provider's profitability by loss of revenue or unnecessary increases in cost.

The design of the questionnaire will also depend on the mode of implementation. The three main methods of questionnaire implementation are:

- postal questionnaire;
- telephone interview;
- person-to-person interview.

This is covered in detail in Section 5.5.

The perception ratings for each implementation methodology must have attached an associated confidence weighting so that the service provider and customer can have a clearer understanding of the data and their interpretation. The design of questionnaires and their implementation are specialised areas, and for maximum benefit should be carried out by professionals with the relevant expertise who also have an understanding of the telecommunications industry.

(e) Publication of results

Only the results of customer surveys specifically obtained for external use are suitable for external publication. These results are best published in a graphical form. A hypothetical example is shown in Figure 8.8.

The information in Figure 8.8 shows the variance of customer perception of voice quality over time. From this information a service provider can establish an acceptable level of customer perception (e.g. 80% of customers think that voice quality is good or better). This type of information can also be used to trigger action by the service provider. For example, the service

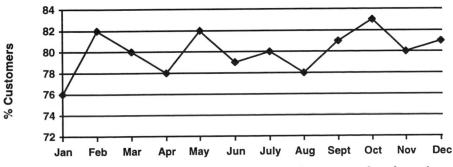

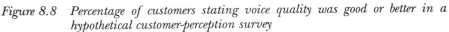

Figure 8.8 *Percentage of customers stating voice quality was good or better in a hypothetical customer-perception survey*

provider can assign limits to this parameter and if the perception falls within those limits no action will be taken. However, if the perception falls outside those limits, action should be taken to correct the situation.

8.4.3 Perceived quality variance analysis

One of the main activities in the management of customers' perception of quality is the correlation of customers' ratings of expected quality and experienced quality, as shown in Figure 8.9.

From Figure 8.9 it can be seen that perceived quality is a function of expected quality and experienced quality. When the expected quality equals the experienced quality, the quality perception is neutral or what was expected and customers will be satisfied (i.e. 1:1 correlation). If the expected quality is greater than the experienced quality, the quality perception will be less than one and customers will be dissatisfied. However, where experienced quality is greater than the expected quality, the quality perception will be greater than one and customers will be satisfied with the quality. But any variance between the perceived quality and the expected quality (i.e. gap 2), should give rise to further investigation depending on the size and magnitude of the variance. For example, if the variance is positive (i.e. expected is smaller than the experienced), the service provider could gain additional revenue and market share by increasing customer expectations without any detrimental effect on customer perception. However, if the variance is negative (i.e. expected is greater than experienced), the service provider must initiate action to correct the problem or suffer the inevitable loss of customers and revenue.

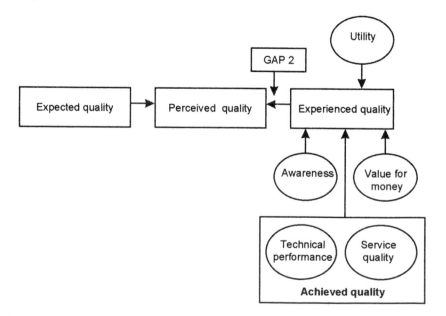

Figure 8.9   Perceived-quality model

To make valid comparisons to determine perceived quality it is necessary to compare like with like. For example, using the criteria in Figure 8.7, customers were asked to rate voice 'call quality'. It can be seen that if customers rated their expected 'call quality' as 5 and their experienced 'call quality' as 4, then, from the above, it is clear that customers are dissatisfied with voice 'call quality'. Where customers are dissatisfied with an aspect of quality, as in this example, the service provider will need to implement a quality-improvement programme, but before any such programme is instigated attention should be paid to the following points:

- allow for the subjective element within customers' responses;
- check survey validity; and
- check customer priority rating of the quality element.

Subjective element of customer rating

The principal shortcoming of any customer rating is its sensitivity to subjectivity. In managing the subjectivity of a customer's perception of quality the following points should be considered:

Overcoming unjustified bias in customer awareness: In Section 8.2.2 was mentioned that customer awareness could affect their perception of quality. Where such perceptions are obviously coloured or based on inaccurate information, the customers' awareness needs to be raised to overcome this problem. For example, customers perceive that the service provider's billing formats do not meet their requirements but are unaware of the range of billing formats on offer and the customisation available within those standard formats. Then the customers' perception can be raised significantly by a simple customer-awareness programme, communicating the true number of billing formats available and also the customisation of those formats.

Management of customer perception by advertising: In Section 8.2.2 was mentioned the relationship between advertising and customers' perception of quality, showing that advertising can be used to improve perception of a product, service or service-provider image. However, care must be taken when using advertising to promote perception because the perceived quality could remain low or even deteriorate if the advertising promises too much.

Survey validity

Before an attempt is made to implement any quality-improvement programme, the validity of the survey needs to be checked or reviewed, as some of the results or areas for improvement may have not been well defined within the original survey. This review should check the customer selection, sample size, questionnaire design and method of implementation. Audit programs are likely to highlight any shortcomings with a perception study, and regular audits should be applied to customer surveys.

Customer priority

Before committing valuable resource to a quality-improvement programme, it is advisable to gauge the customers' priority weighting for the quality element to be enhanced. For example, if a perception survey showed that customers were dissatisfied with 'voice-call quality', but their priority weighting for this parameter was low, then the service provider would do better to use resources to improve a different parameter with a higher customer-priority rating.

8.4.4 Modification to the quality programme

In quality management, any modifications to the activities influencing quality must be evaluated before implementation. To illustrate, if the perception rating on availability of a particular service has declined and the cost of improvement far outweighs the likely revenue increase, it may not be worthwhile for the service provider to carry out the modifications. However, other considerations may apply. It may be in the interests of the service provider to spend the necessary amount to improve its company profile on quality. Such profile improvement could have a beneficial effect on the service provider's entire portfolio.

The improvements indicated from the perception ratings could also be submitted to the customers'-requirements capture for the next round of the management activities of the quality cycle.

8.5 Conclusions

Customer's perception of quality, expressed by opinion ratings, must be used for effective management of quality by the service providers. In the future, more use could be made of customer-perception surveys as part of the decision-making processes within a service provider's organisation. More work is also likely to be carried out in understanding the relationships between customer perception, expected quality, delivered quality, competition, technology, advertising, customer requirements and other variables. Improvements identified from perception-survey ratings could be incorporated in the customers'-requirements capture process for inclusion in the next round of the management activities of the quality cycle, thus completing the quality cycle.

8.6 References

1   BOLTON, R. N., and DREW, J. H.: 'A longitudinal analysis of the impact of service change on customer attitudes', *J. Marketing*, 1991, **55**, pp. 1–9
2   BROWN, Tony: 'Understanding BS 5750 and other quality systems' (Gower Press, 1993), section 3.6

3   MULLEE, A. W., and FAULKNER, R.: 'Planning for a customer responsive network'. *Sixth international network planning symposium*, September 1994, pp. 337–351
4   ZEITHAML, V. A., PARASURAMAN, A., and BERRY, L. L.: 'Delivering quality service' (Free Press, 1990)
5   GRONROOS, C.: 'Service management and marketing – managing the moment of truth in service competition' (Lexington Books, Mass., 1990) pp.41–47
6   GOTLIEB, J. B., and SAREL, D.: 'The influence of type of advertisement, price and source credibility on perceived quality', *J. Acad. Marketing Sci.*, 1992, **20**, (3), pp. 253–260

Chapter 9

Service surround and service management

9.1 Introduction

A major factor in a customer's choice of service provider will be the quality or level of customer service. For example, many people shop at certain stores, stay at certain hotels, eat in a particular restaurant or take their car to a particular garage based on the level of customer service provided by that establishment [1]. This aspect of customer service encompasses such factors [2] as:

Reliability: the ability to provide what is promised, dependably and accurately;
Responsiveness: the willingness to help customers and provide prompt service;
Empathy: the caring and individual attention provided to customers;
Assurance: the knowledge and friendliness of employees and their ability to convey trust and confidence; and
Flexibility: the willingness of employees to help customers meet their needs outside the normal standard processes.

These and other similar factors are often referred to as the 'soft issues' of quality and are of very important within the area of customer service. These features, will be crucial in terms of a service provider's shareholder value and thereby corporate success. Put simply, increasing shareholder value through profit growth is achieved by adding value to the basic product. Value is about both price and service quality. Service† has a crucial role to play if customers are to make value judgements and not just price judgements.

The need to differentiate on service quality in telecommunications has already been recognised in the US where MCI, which historically concentrated on price, has changed its focus to service quality. Within the UK, AT&T has announced that it will not seek to compete on price as a market-entry strategy, but that the battle will be on advanced services, service surround and service management. Service surround is based around the way a customer interacts or does business with a service provider and the customer's experience and perception of that interaction. Service surround is more than the technical

---

† The term 'service' in this chapter refers to pure service (e.g. 24 hour access to customer service, one-stop shopping, professionalism of service providers etc.) and not a telecommunication product or service (e.g. telephone service or frame-relay service).

features of a product or service, it is also more than the specific service wrapped around any individual product or service, and includes every single interaction with the service provided, whether product related or not.

What does moving from a commodity to a service brand position mean for service providers? It means two things. First, the service provider needs to deliver excellent service value. Secondly, the service provider must communicate this to its customers so that they believe it, are satisfied with it and hence remain loyal throughout their lifetime. This basic model is shown in Figure 9.1.

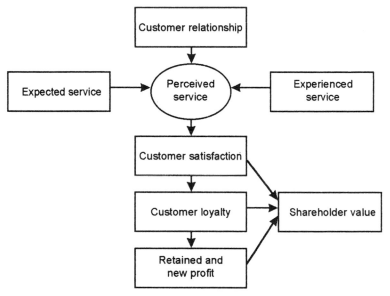

Figure 9.1   Service model

In an industry where technology and customer needs are changing at a rapid rate, the only constant will be the customer relationship. The main tangible element of this relationship is customer service and the way in which customers perceive that service. It is the crucial factor that the service provider must get right to develop a true service ethos; it must move from one-off transactions through interactive dialogue and to the ultimate goal of developing a lasting customer relationship built on customer-service quality.

Within the telecommunications industry, customers are turning more towards this concept of customer-service quality as a differentiation factor which covers the total customer experience with the supplier, and moving away from the pure product performance or product-specific service performance towards customer service, as can be seen from Table 9.1.

Table 9.1 shows that customers are becoming more concerned with service issues in addition to the traditional technical performance of any specific product. However, some may argue that product-specific performance and customer service are one and the same. This would be true if the service provider were a one-product supplier. However, service providers will usually have a portfolio of products, and customers will expect a generic level of service across the entire portfolio irrespective of which product is being consumed. This concept is illustrated in Figure 9.2.

Table 9.1   Customers' changing service requirements (source: developed from internal studies on QoS)

Customers' top-ten communication requirements (yesterday)	Customers' top-ten communication requirements (today)
1 Loss of service	1 Access to service support
2 Frequency of loss	2 Availability
3 Line reliability	3 Professionalism of staff
4 Agreed installation dates	4 Service flexibility
5 Efficiency of engineers	5 Response time for customer assistance
6 Clarity of voice lines	6 Time to repair
7 Clarity of nonvoice lines	7 Repair carried out right first time
8 Ease of billing verification	8 Repair update
9 Repeat faults	9 Time for provision
10 Update on fault repairs	10 Provision flexibility

From Figure 9.2 it can be seen that customer service has two distinct parts: 'generic service' with standard performance levels across a service provider's entire portfolio, and 'service specific' with differing levels of service performance depending on the specific product. For example, reliability, i.e. the ability to provide what is promised dependably and accurately is a generic service attribute. Under this service attribute will be a number of service characteristics (e.g. access to a service centre 24 hours per day 365 days a year,

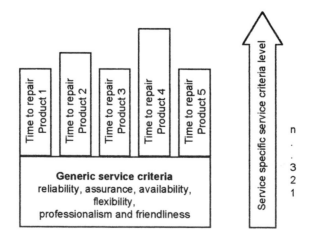

Figure 9.2   Generic and product specific service

all calls answered within 30 seconds etc.), which would have the same performance target across the service provider's entire portfolio irrespective of the product involved (i.e. product independent). However, product-specific service performance will vary from product to product. For example, the service feature 'time to repair', will vary from product to product depending on the complexity of the product, the market segment and the price, e.g. telephony 'time to repair' for the residential market is three working days, but for the business market is three hours.

This holistic approach to customer-service quality, covering generic customer service, product-specific customer service and product perform-ance is termed 'the service surround'.

9.2 Service-surround concept

The service-surround concept is based on the way a customer interacts or does business with the service provider and the customer's experience of that interaction (Figure 9.3). This interaction need not be product specific, e.g. a customer or potential customer may complain about an employee's attitude or state of dress. This customer or potential customer will be interacting with the service provider within the service surround layer shown in Figure 9.3, and the customer or potential customer's perception of the service provider's 'customer service' will be determined by the outcome of that interaction.

As can be seen from Figure 9.3, the customer's service experience takes place at three levels within the service provider's organisation: the product level, product service-specific level and at the service-surround level.

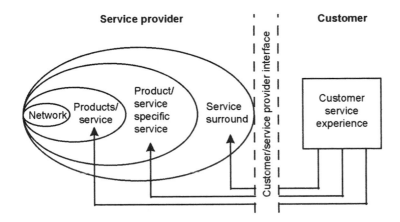

Figure 9.3   Service-surround concept

9.2.1 Product/service level

Within the telecommunications industry, most interactions at the product level take place automatically. For example, when a customer wishes to make a telephone call, the customer picks up the telephone, keys the required number, gets through to the required destination, completes the transaction (telephone call) and hangs up. The quality of the experience at this level will be based purely on the product's performance throughout the experience. The performance criteria will normally consist of the product features, their durability and the technology capability available during that interaction. Figure 9.4 shows a model (described in Chapter 4) which enables a comprehensive set of performance characteristics for this type of interaction to be derived. For example, 'speed of connection' is time taken from the instant dialling is complete, to the instant the signal is received from the network to indicate the status of the called party, and 'security of information transfer' means that the telephone call cannot be overheard by a third party.

9.2.2 Product/service-specific service level

The quality of the experience at the product-specific service level will be based on an individual product and the service associated with that product. For example, before a product or service can be used, it must be purchased and provided, and a customer's perception of the product will start to form

Service function \ Service quality criteria	Speed 1	Accuracy 2	Availability 3	Reliability 4	Security 5	Simplicity 6	Flexibility 7
Service management — sales & precontract activities 1		X					
provision 2				X			
alteration 3							
service support 4							
repair 5							
cessation 6							
Connection quality — connection establishment 7	X						
information transfer 8					X		
connection release 9							
Charging & billing 10							
Network/service management by customer 11							

Figure 9.4   Performance characteristics at product and service level

based on that product's specific service, even before normal use can take place. Therefore, a customer's perception (perception = experienced performance – expected performance) will be based on a number of performance features, as shown in Figure 9.4. For example, 'accuracy of sales' means all sales information is accurate, i.e. product/service features, performance and price are as stated in presales literature and subsequent negotiation. 'Reliability of provision' means that the product or service was installed as agreed, on time with all the features available, and 'billing simplicity' means that the bill is easily interpreted and easily checked.

9.2.3 Service-surround level

The quality of the experience at the service-surround service level will be the sum of the other two levels plus additional service features which are non-product or nonproduct-service specific and are common to the service provider's entire portfolio, including company image. For example, a customer or potential customer makes an enquiry which is not product or product-specific service related; whether the service provider's staff are responsive, flexible, knowledgeable and friendly will impact on the customer or potential customer's experience. If the experience is good, the customer or potential customer will have a positive reaction to the service provider, but if the experience was bad then the customer or potential customer's reaction will be negative. This experience could turn a potential customer into a real customer. The service provider must adopt this holistic approach to service quality and consider customer perception throughout the entire life-cycle of its entire portfolio, coupled with the company and brand image, and not just consider any particular product specification. This totality of customer experience takes place across the customer/service-provider interface.

9.2.4 Customer/service-provider interface

A customer's experience of a telecommunications product or service normally takes place across a customer/service-provider interface, but during normal use this interface may be transparent to the customer. For example, whenever a customer makes a telephone call, this interaction takes place across an interface which consists of the telephone network. Many such interactions of telecommunications use can take place across this interface with the only customer/service-provider contact being the bill payment. However, if an interruption to this normal use occurs, for example, a request for additional product features, new-product provision, a problem or a general enquiry, then the interface process changes to one of human intervention through the service surround. Hence, the quality of the experience will now be subject to a whole host of new criteria as discussed earlier.

   If all product and service interactions take place across the customer/service-provider interface, then the challenge for the service provider is to manage the processes and interactions across this interface successfully. This activity is termed 'service management'. Therefore, a customer's perception of a service provider will be dependent on the service-provider's service management.

9.3 Service management

9.3.1 General

Service management from the service provider's perspective is the management of the customers' total experience with the company. In practice, this means managing the total process and management of the tasks produced by interactions across the customer/service-provider interface (see Figure 9.5).

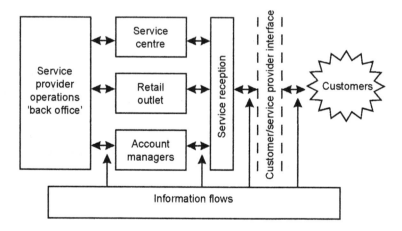

Figure 9.5   Service-management model

9.3.2 Service reception

Service reception is the name given to the generic 'single point of contact' within a service provider's organisation for customers or potential customers, and it is usually the first point of contact between a customer and a service provider. The service reception will in practice normally include one or more of the three main function areas of 'retail outlet', 'service centre' and 'account management', depending on the customer, as shown in Figure 9.5. For example, a business customer may obtain access to the service provider's organisation via a business service centre or very large customers may have a dedicated account manager; residential customers will normally access via a high-street retail outlet or by telephone via a residential service centre. Irrespective of customer type or segment, the service reception will reflect the service provider's attitude to customer service; therefore, customer handling within the service reception will be a key element in a customer's perception of a service provider.

Detailed customer requirements of a service reception are either not well understood or not effectively implemented by the majority of service providers. However, for customers to have a positive perception of customer service reception, the following key areas will need to be addressed:

- simplification and convenience of contact for all customers (i.e. telephone, facsimile, e-mail, letter and direct system access/electronic access);
- contact that builds relationships and thus loyalty;
- quick response to customer contact;
- keeping customer commitments;
- proactive feedback on provision, fault or complaint progress at the frequency and by whatever means the customer specifies;
- no more than one customer 'hand off', i.e. customers should not be transferred more than once before completion of their enquiry;
- people with the skills, competence and information to deal with the vast majority of enquiries with no 'hand off';
- access to function experts (i.e. product managers) for complex enquiries;
- consistency when passing customers to other parts of the organisation; and
- 'sign-off' with each customer for each enquiry completion.

These requirements are consistent with the findings from the Henley Centre report on Teleculture 2000 in which the three Cs (Convenience, Cordiality, and Consistency) emerged as being very important in terms of customers' experience of service within the telecommunications industry. The implications of these are illustrated in Figure 9.6.

If service providers are to achieve excellent customer service quality and high customer satisfaction, they will need to focus on a user-friendly highly responsive customer reception. Totally satisfied customers are six times more likely to repurchase over the next six months than are merely satisfied customers [3].

9.3.3 Service centres

Service centres are the main point of contact between service providers and their business customers. Service centres usually provide a 'one-stop' shop for most business customers and are equipped with the necessary resources to solve most business customers' communication needs, from enquiries and sales through to repair and complaints handling.

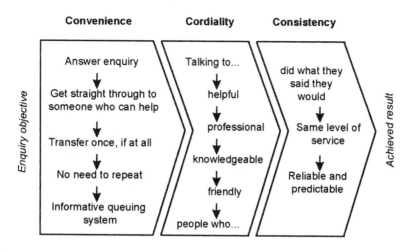

Figure 9.6   Three Cs of customer service experience

9.3.4 Account management

A service provider's largest customers will normally have access through account management to dedicated account managers. Account managers are highly trained individuals with a in-depth knowledge of the business of a specific customer or group of customers (e.g. in-depth knowledge of the travel industry) and their business communication needs. They will ensure that the service providers' largest customers have customised attention and will personally ensure that all their needs are satisfied.

9.3.5 Retail outlets

These are normal high-street access points geared to cater for the service provider's small business or personal customers. This type of access offers customers the opportunity to interact with the service provider at their leisure during normal retail opening hours. Customers will have access to trained staff with a good overview of all the service provider's portfolio of products and services and their applications, with secondary instant access to a service centre with in-depth expert knowledge of all product- and service-portfolio applications. The retail outlet will be able to meet the entire range of customer needs from enquiries and sales through to repair and complaints handling.

9.3.6 Service-provider operations

The service provider's operations normally take place in a back-office environment and this is the place where customers' needs are translated into commercially viable products and services. In effect the back office is the factory encompassing all the activity necessary to deliver service to the service provider's customers.

9.4 Service management process

9.4.1 Components of service-management process

Figure 9.5 shows the service-management model and the flow of tasks and information across the customer/service-provider interface and through the service provider's organisation. This clearly shows the service-management function to be the management of all activities across the customer interface. The major tasks produced by this activity are termed service-management components and the major components of service management are:

- enquiries;
- provision of service;
- repair of service;
- billing of the service; and
- customer complaints.

9.4.2 Enquiries

'Enquiries' is an all-embracing element of service management and covers any interaction not covered by the other four components. Customers may require information about products and services, availability, delivery dates and prices. They may require information on progress, e.g. on what is happening with their order or fault. They may require information about the content of their bill or wish to know the procedure for disputing a bill. The answers to all these enquiries should be obtained by the process which supports the other four service-management components.

The enquiries component of service for a service provider also includes all the tasks associated with 'operator services'. The types of operator services offered include directory enquiries and emergency services. A large service provider should ensure that the customer-facing staff have the aptitude to handle customers sensitively and provide a measure of understanding and patience in dealing with the customers. One of the principal contributory factors in the formation of opinion among the public towards the service provider is through the service offered by its operators. In modern telecommunications systems, most of the switching is carried out automatically. When an operator is sought, it is usually for a service which requires personal attention, and the customer is fairly sensitive to the treatment received during the transaction of these types of services.

9.4.3 Provision of service

The 'provision-of-service' process requires the service provider to undertake a series of tasks which may vary from product to product and service to service. However, the process should be consistent from the customer's perspective, irrespective of which product or service is being provided. The generic process for service provision will follow the stages shown in Table 9.2.

Table 9.2   Service provision

Stage	Task	Comments
1	Identify the customer	Is this an existing customer or is there need to gather data from a new customer?
2	Establish customer requirements	Whether they wish to place an order, amend an existing order, cancel an existing order or simply make an enquiry. Enquiries about products and services also include descriptions, suitability, availability, delivery dates and pricing information. Enquiries can also be made about the progress of an existing order. Progress enquiries will require access to the 'manage-work' process of the provision-service component
3	Accept orders	For products and services, checking availability, pricing products and services, gathering customer data for operational, financial, marketing, product and service aspects, committing to a delivery date, allocating appointment slots, job prioritisation and job breakdown into activities. Amendments and cancellations are also included within this process
4	Update inventory	Where capacity has been used for service provision, the appropriate inventories, including the customer inventory, need to be updated.
5	Manage work	For scheduling and allocation of activities to the appropriate skilled resources. Despatch of people for task execution. Progress monitoring and jeopardy warnings. Resource recording against task completion and aggregation of task compilations to activity and job completion
6	Test	To ensure service quality prior to handover to the customer
7	Handover to customer	Close order for billing purposes
8	Bill customer	Process debits and credits, produce bill and despatch to customer

9.4.4 Repair service

The 'repair-service' tasks will also vary from product to product and service to service. However, the process must also be consistent from the customer's perspective, irrespective of which product or service is being repaired. The generic process for 'repair' will follow the stages illustrated in Table 9.3.

Table 9.3 Repair service

Stage	Task	Comments
1	Identify the customer	Is this an existing customer or is there need to gather data from a new customer?
2	Establish customer requirements	Whether they wish to report a fault or make an enquiry. Enquiries in this instance are usually about the progress on repairing a fault. Progress enquiries will require access to the 'manage-work' process of the repair-service component
3	Accept fault reports	On products or services, gathering customer data for operational, financial, marketing and products and services aspects, job prioritisation and job breakdown into activities
4	Test	To verify fault and cause
5	Manage work	For scheduling and allocation of activities to the appropriate skilled resources. Despatch of people for task execution. Progress monitoring and jeopardy warnings. Resource recording against task completion and aggregation of task completion's to activity and job completion
6	Update inventory	Where spare capacity has been used to overcome a fault, including the customer inventory
7	Test	To ensure service quality prior to handover to the customer
8	Handover to customer	Close fault report

9.4.5 Billing service

As with the 'provision service' and 'repair service', the 'billing-service' process must also be generic from the customer perspective, irrespective of the product or service in question. Billing interactions may concern individual products, services or the bundling of a number of products and services, but still the process across the customer/service supplier interface must be uniform. The generic process for the billing service is shown in Table 9.4.

Table 9.4   Billing service

Stage	Task	Comments
1	Identify the customer	Is this an existing customer or is there need to gather data from a new customer?
2	Establish customer requirements	Whether there is an error in the bill which needs amendment or this is an enquiry. Enquiries may be about bill content, presentation, calculation and frequency
3	Check customer bill	To establish whether the contents are correctly presented and the customer-billing requirements have been meet. If there is a mismatch, record the amendment
4	Check customer inventory	To establish whether the inventory recorded on the bill correlates with that recorded against the customer and that they both correlate with what the customer actually possesses in terms of products and services. If there is a mismatch, record the amendment
5	Check customer usage	To establish whether the usage information recorded on the bill correlates with that recorded against the customer and they both correlate with what the customer actually used. If there is a mismatch, record the amendment
6	Accept bill amendment	Apply the amendments to the customer account
7	Update inventory	Where appropriate including all customer records
8	Handover to customer	Close amendment for billing purposes.
9	Bill customer	Process debits and credits, produce bill and despatch to customer

9.4.6 Customer complaints

Many customers' choice of a service provider is based not only on the service provided at the time of sale, but also on the expected after-sales service (i.e. what happens if something goes wrong). Therefore, a key area of service management affecting customers' choice of supplier will be how that service provider responds to complaints [4].

Complaints can be made by customers on wide range of issues, covering all areas of the product and the service surround from rudeness of staff, lack of punctuality, broken promises through to noisy telephone lines.

It takes a long time before a 'normal' customer makes a complaint. Usually, a customer will put up with deviances from expected performance, provided that these are few and tolerable. However, when such deviances become intolerable, customer's complain. The customers' attitudes towards complaining will depend on their determination to seek redress when dissatisfied [5,6,7]. Some customers are assertive and seek redress whenever they are dissatisfied, while others are reluctant to complain no matter how dissatisfied they become. Nevertheless, complaints have a detrimental effect on a service provider. It has been said that a dissatisfied customer will tell nine people and satisfied customer will probably tell only one. Notwithstanding the precise figures, it is certain that a dissatisfied customer will spread bad feeling about the service provider. Where there is choice, a customer may switch brands. When this happens it can take seven to ten times more resource to regain a customer than to maintain one.

Therefore, this aspect of service management is critical for the long-term profitability of a service provider. Firms which develop a reputation for rectifying customer complaints successfully, are likely to develop customer loyalty and, over time, increase market share, profitability and hence shareholder value.

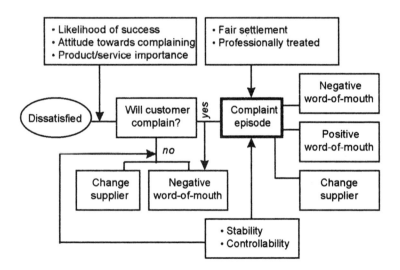

Figure 9.7   Complaints process [1]

When customers are dissatisfied, they will respond in a number of ways depending on the likelihood of success. Their personal attitude to complaining and the importance of the product or service to them is shown in Figure 9.7. Dissatisfied customers will choose whether to complain or not based on these criteria. Customers which choose not to complain may either change supplier or engage in negative word-of-mouth comment (i.e. tell other customers of their dissatisfaction). Customers which decide to complain, depending on the outcome of their complaint (e.g. fair settlement) and how they were treated during the complaint process, will either enter into positive word-of-mouth comment or negative word-of-mouth comment or change supplier. Also, complainants which believe that the problem was controllable (i.e. could have been avoided) or stable (i.e. similar problems will occur in the future) are more likely to change supplier or warn other customers not to do business with that supplier [8]. It is vitally important for the supplier to empathise with the customer, assure the customer that the problem will not occur again and always offer a sincere apology [9].

Management of complaints
The management of complaints is simple. With the application of the will, complaints can be overcome very quickly. A model for the handling of complaints is illustrated in Figure 9.8. The first two boxes in Figure 9.8 'receive, log and acknowledge complaint' and 'resolution and advise customers' follow the same process as the other major components of service management, as shown in Table 9.5.

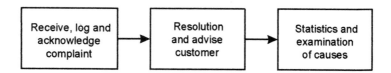

Figure 9.8   A methodology for processing complaints

Reception and logging of complaints
Wide publicity should be given to the methods of making complaints. The customers should be able to complain directly to the following:

● the head of the service provider's organisation (i.e. chairman's office);
● the association which represents their interests, e.g. Telecommunications User's Association;
● the regulator;
● elected government representative, if the service provider is fully or partly owned by the government; and
● the service provider's customer reception (Figure 9.5).

Table 9.5   Complaint process

Stage	Task	Comments
1	Identify the customer	Is this an existing customer or is there a need to gather data from a new customer?
2	Establish customer requirements	Whether they wish to complain or make an enquiry. The enquiries may be about the complaints process
3	Accept complaint	Gather all relevant information, log complaint. Assure/ advise the customer of complaint process, prioritise complaint and breakdown into activities and inform customer of complaint-resolution plan
4	Manage work	For scheduling and allocation of activities to the appropriate skilled resources. Despatch of people for task execution. Progress monitoring and jeopardy warnings. Resource recording against task completion and aggregation of task completions to activity and job completion. Keeping customer regularly informed of progress
5	Test	To ensure service quality prior to handover to the customer
6	Handover to customer	Close complaint

Customers should be able to make complaints in writing, by phone or by any electronic means. When a complaint is made, it should be logged and acknowledged by the service provider. The log should contain the date of complaint and facilities for further information such as the progression of the complaint until resolution. All complaints should be logged onto a computer, irrespective of the method of arrival at the service provider's organisation. Customers should be given a reference number and should not have to quote the complaint all over again should a different person answer a follow-up query from the customer.

Resolution of complaints

Service providers must understand that customers have reached the end of their patience when making a complaint. The wording in an Australian report on complaints handling [10] describes well the warning to service providers. It says: 'poor treatment of complaints by a monopoly is inexcusable; in a competitive situation it can be downright foolish'.

At the end of the resolution of a complaint, it is highly recommended for the service provider to make a sincere apology and assure the customer that the problem will not occur again, and in certain circumstances make some amends for the inconvenience suffered.

Statistics and examination of causes

Production of statistics on the number of complaints, their nature and the time for resolution could have major benefits for the service provider in several ways:

- for publication, as required by the regulator, customer user groups and individual customers, showing the service provider's commitment to customer service and measured performance;
- for the internal management of quality by the service provider; the service provider could study the complaints profile to look for a pattern and identify the underlying causes with a view to problem eradication;
- to gain competitive advantage through benchmarking complaints resolution of other service providers; and
- enhancement of company and brand image.

9.5 Service-management systems

In many service provider's organisations, the processes which underpin the components of service management have been automated to varying degrees. These components of service management can be viewed as the service providers' 'production lines' and the automated process the 'production system'.

Tables 9.2 – 9.5 show that the front-end processes of 'identify the customer' and 'establish customer requirements' are common to all the major service-management components. It therefore follows that the systems which support these initial processes form the basis of the service provider's customer-reception system. Because the fundamental process are the same, they should be treated as a common system which shares common data. The data needed relate to customers and products and services offered by the service provider. Therefore, only two major databases are required for customer reception across the full range of service-management components and across the entire portfolio of products and services, backed up by smaller systems allowing specific tasks to be performed and monitored.

The requirements of service-management systems of the future are inextricably linked to the service provider's strategy. For example, if the service provider adopts a reception strategy which is tailored to provide service to specific customer groups across the entire service provider's portfolio of products and services, with global geographic coverage, it follows that the support systems for service management must provide a matching capability.

9.6 Service-management people

The quality of a service provider's people is of paramount importance if the organisation is to operate effectively. All businesses are people businesses, and nothing can get done without human interaction of some kind. While technology is a key enabler, success will ultimately be determined by the people who operate the systems. This means having the right numbers of people available at the right times, and having people with the right skills and culture. Within the telecommunications industry most of the human contact with customers will be via the service-management processes. Therefore, the service-management people have a major role to play in the customers' perception of the quality of the service provided. Customers expect a service provider's service-management people to be knowledgeable about the portfolio of products and services, friendly, flexible and resourceful in meeting their needs and professional in all their dealings.

If service providers are to achieve service excellence and the rewards of retaining satisfied customers and gain new profit and hence increased shareholder value, they need to look very carefully at the calibre of their service-management people. With the correct company culture, and together with 'empowered' and motivated people with the correct tools, the service provider has all the necessary ingredients to be successful.

9.7 Service surround and service-management performance

To assess the effectiveness of the service surround and the service provider's service-management capability, the components which make the service-management functions (e.g. enquiries, provision of service, repair of service, billing of the service and customer complaints) must be measured against customer requirements and benchmarked against other service provider's performance.

Christian Gronroos listed six major criteria for good service quality [11]:

Professionalism and skills: The customers realise that the service provider, its employees, operational systems and physical resources have the knowledge and skills required to solve their problems in a professional way (outcome-related criteria).

Attitudes and behaviour: The customers feel that the contact persons are concerned about them and genuinely interested in solving their problems in a friendly and spontaneous way (process-related criteria).

Accessibility and flexibility: The customers feel that the service provider, its location, operating hours, employees and operational systems are designed and operate so that it is easy to gain access to the service and that it is prepared to adjust to the demands and wishes of the customer in a flexible way (process-related criteria).

Reliability and trustworthiness: The customers know that, whatever takes place or has been agreed on, they can rely on the service provider, its employees and systems to keep promises and perform with the best interests of the customers at heart (process-related criteria).

Recovery: The customers realise that, whenever something goes wrong or something unpredictable unexpectedly happens, the service provider will immediately and actively take corrective action (process-related criteria).

Reputation and credibility: The customers believe that the operations of the service provider can be trusted and that it stands for good performance and values which can be shared by them (image-related criteria) [9].

Gronroos concludes that the six criteria can be used as a framework on which to base service quality measures.

These and other criteria developed by Zeithaml, Parasuraman and Berry [12] for measuring service quality termed 'SERVQUAL' (Section 8.4.2) form a comprehensive framework from which to develop specific customer-orientated measures of service management and hence measure the service surround.

The physical measures will naturally fall into two categories: quantitative (for example, the total number of complaints successfully resolved in one month) and qualitative (for example, the customer perception of the service provider's employees' professionalism). The results from both sets of measures can then be used by the service provider to gain and sustain differentiation and competitive advantage through service. This establishes a reputation for leading with service followed by products to deliver customer solutions which meet all their requirements.

9.8 Conclusions

A major factor in a customer's choice of service provider is that service provider's level of service quality. Then the service surround and the management of that service surround (i.e. service management) will become one of the main battlegrounds where service providers attempt to differentiate themselves from their competitors. In order to achieve this differentiation, service providers will need a thorough understanding of the service-surround concept, the customer interactions within this service surround through the customer/service-provider interface and hence the management of these interactions (service management).

If service management is to be one the main battle areas, the choice of measurements used by the service provider to prove its competence in service management will be vitally important. The measures must be representative of the service-management components and the processes used to deliver service management and also be customer orientated, reflecting the customers' service-management priorities.

Therefore, service providers wishing to differentiate themselves from the competition and gain competitive advantage must refocus their efforts on customer-orientated service management.

9.9 References

1 BLODGETT, J. G., WAKEFIELD, K. L., and BARNES, J. H.: 'The effects of consumer service on consumer complaining behaviour', *J. Service Marketing*, 1995, **9**, (4), pp. 31–40
2 KERIN, R. A., JAIN, A., and HOWARD, D. J.: 'Store shopping experience and consumer price-quality-value perceptions', *J. Retailing*, 1992, **69**, (Winter), pp. 376–397
3 JONES, O., EARL, W., and SASSIER, JUN.: 'Why dissatisfied customers defect', *Harvard Business Rev.*, Nov.–Dec. 1995
4 GOODWIN, C., and ROSS, I.: 'Consumer evaluations of response to complaints: whats fair and why', *J.Service Marketing*, 1990, **4**, (Summer), pp. 53–61
5 BEARDEN, W. O., and MASON, J. B.: 'An investigation of influences on customer complaint reports', *in* KINNEAR, T. C. (Ed.): 'Advances in consumer research (Association for Consumer Research, Provo, UT, 1994)
6 RICHINS, M. L.: 'A multivariate analysis of responses to dissatisfaction', *J. Acad. Marketing Sci.*, 1994, **15**, (Fall), pp. 24–31
7 SINGH, J.: Exit, voice and negative word-of-mouth behaviour: an investigation across three service categories', *J. Acad. of Marketing Sci.*, 1990, **18**, pp. 1–15
8 FOLKES, V. S.: 'Consumer reactions to product failure: an attributional approach', *J. Consumer Res.*, 1994, **10**, (March), pp. 399–409
9 BIES, R. J., and SHAPIRO, D. L.: 'International fairness judgements: the influence of casual accounts', *Social Justice Res.*, 1997, **1**, pp 199–219
10 'Telecom's Handling of Customer Complaints'. Report from the House of Representatives Standing Committee on Transport, Communication and Infrastructure (Australian Government Publishing Service, Canberra, 1991)
11 GRONROOS, C.: 'Service management and marketing — managing the moment of truth in service competition' (Lexington Books, Mass., 1990) pp. 41–47
12 ZEITHAML, V. A.: PARASURAMAN, A., and BERRY, L. L.: 'Delivering quality service' (The Free Press, 1990)

Chapter 10

Network performance engineering

10.1 Introduction

Network performance provides the technical standard of the network in terms of the performance of particular combinations of items of equipment employed to provide services. It is a major contribution to the QoS which customers experience when using the network. It is measured by parameters meaningful to network operators and there is no absolute relationship to QoS. In some cases, several performance parameters may contribute to the same impairment perceived by customers. Also, there are particular key parameters which fundamentally effect the quality of each service offered by the network, e.g. error performance for data services and delay for voice services.

ITU–T *Recommendation I.350* [1] defines network performance as 'the ability of a network (or network portion) to provide functions related to communication between users'. It also comments that network performance should be defined and measured in terms of parameters meaningful to the network provider and independent of terminal performance and user action. The ITU considers the usage of a service in chronological order [2]. So a single connection-oriented call has an *access* phase, an *information-transfer* phase and a *disengagement* phase. It considers the performance requirements for each phase by means of the matrix shown in Figure 10.1. The parameters in the cells are known as *primary* or directly observable and those not directly measured, but observed from the primary parameters, are known as *secondary* parameters.

The access phase encompasses user indication of the required service, the call destination and any required service parameters such as bandwidth requirement, charge advice, closed user group etc. The performance can be measured in terms of speed of call establishment (delay to dial tone and delay to ringing tone), accuracy (no misrouted calls) and dependability (low failure due to congestion).

Information transfer covers the period between call establishment and cleardown. Speed is measured as information-transfer rate, e.g. for data and fax transfer. Accuracy is measured by such things as error rate and distortion, and dependability e.g. the incidence of poor accuracy and low speed transfer.

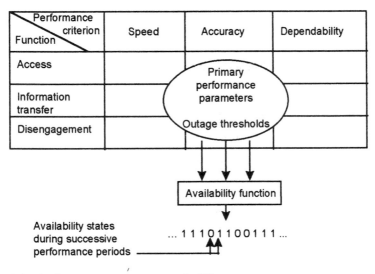

Figure 10.1   Performance-parameter matrix [1]

Disengagement concludes the information transfer and marks completion of the call; hence, speed of cleardown, accuracy of billing and dependability in terms of no premature release or failure to clear are performance parameters.

The mapping of performance parameters, grouped in the three 'ITU' phases of the communication function, on their customer-recognisable QoS parameters is illustrated in Table 10.1.

Performance impairments affecting switched services can be categorised as:

(a) Call-processing performance, which relates to the ability of a network to accept and interpret routeing information from the customer and establish a connection to the required destination within prescribed response times. The main drivers are the performances of exchanges and signalling systems.

(b) Transmission performance is the ability to transport information between source and destination without distortion or undue loss, and is mainly influenced by the performance of transmission systems, although digital switches introduce delay.

(c) Availability performance may be broadly defined as the proportion of time for which satisfactory service is given.

Table 10.1   Performance and customer expectation

Function	Customer expectation	Network performance parameters
Access	Negligible faults/ no loss of service	Network availability
	Minimal call-connect failures	Connection-establishment failure resulting in no tone
		Misrouteing of calls
	Rapid call set-up	Delay to dial tone
		Connection-establishment delay
	Connection-establishment failure due to network connection	Grade of service
Information transfer	Good transmission quality	Transmission loss
		Impulsive noise
		Psophometrically weighted noise
		Single-frequency noise
		Loss distortion with frequency
		Group-delay distortion
		Propogation delay
		Echo
		Stability
		Quantising distortion
		Sidetone
		Crosstalk
		Error performance
		Jitter and wander
		Slip rate
Disengagement	No cut-offs	Premature release of established connections

Little is known about the dimensions of QoS as seen by the customers and their relative importance; hence QoS targets are generally set in terms of past experience rather than theoretical considerations. Criteria include billing errors, waiting lists for provision of service, wrong numbers, time to repair, poor transmission, fault incidence, failure of call completion and dial-tone/post-dialling delay. Figure 10.2 shows how these can be structured as a tree of concepts. The most important concepts are:

(i)   availability, i.e. the ability of an item to perform its required function at any given instant;

(ii) trafficability, i.e. the ability of an item to fulfil operational traffic demands under stated conditions of use.

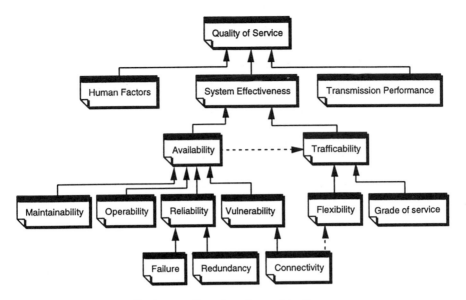

Figure 10.2 Factors affecting quality of service and performance

Network-performance parameters are established, generally taking account of ITU-T recommendations and economic considerations and are used for system design, network planning, specification and procurement of equipment, network operation and service specification. The parameters are often quoted in terms of 'design objectives' which are the values to be expected when equipment is operating in a defined environment and generally used in specification clauses in procurement documents. It is important to distinguish between design objectives and the various commissioning and maintenance values adopted operationally as illustrated in Figure 10.3.

Overall network-performance parameters are generally empirically derived and must be carefully apportioned over the contributing network elements to give the required target result. The apportionment is influenced by economic considerations and the requirement to give the associated end-to-end QoS target to customers for international calls, calls between networks of other network operators and where calls terminate in customer private networks. The application of this principle for performance parameters is shown in Figure 10.4.

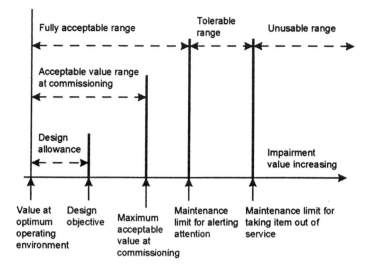

Figure 10.3   Performance objectives

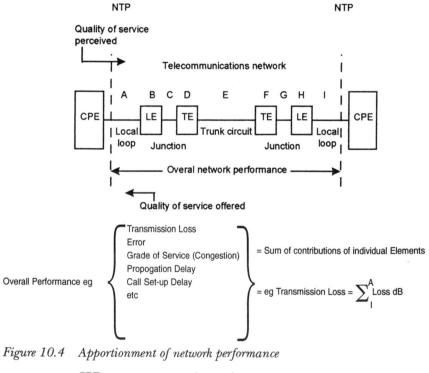

Figure 10.4   Apportionment of network performance

CPE   customer-premise equipment
NTP   network-terminating point
LE   local exchange
TE   trunk exchange

10.2 Performance management

Performance management is an essential element of the business process of 'operating the network'. It is necessary to ensure that planned network performance is achieved so that, in a competitive situation, the dominant telecommunications company can retain, and perhaps gain, market share by differentiating on the QoS delivered to its customers. Such a strategy requires good measurement of the QoS delivered as an essential feedback loop of quality management. Support systems related to performance management fall within the 'performance management' functional area of the Network Control Layer of the ITU specified Telecommunications Management Network (TMN) network-management architecture [3], shown in Figure 10.5.

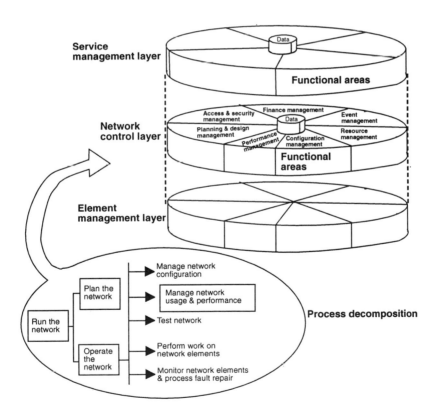

Figure 10.5   Relationship of processes and network-management functional areas

Performance management must:

(a) measure the QoS delivered to customers for each service offered. It must be recognised that this is just one of the four dimensions of service quality, namely: quality delivered, quality offered, quality required and quality perceived. Also, the performance measured only represents those parameters which can be captured from the network and its operations and not the 'softer' dimensions of customer satisfaction;

(b) provide an overall check on the overall health of the network with data to assist the planning and operations process;

(c) review trends in the performance of individual equipment to assist purchasing and design decisions;

(d) measure efficiency, effectiveness and cost of key processes and workstrings;

(e) closely monitor QoS delivered to other licensed operators (OLOs) connected to the network.

10.3 Transmission plan

10.3.1 General

The network transmission plan is a set of network-design guidelines recommended by the ITU-T to ensure that end-to-end transmission is stable and acceptable to users. The recommendations lay down references against which national and international networks can be designed and cover such things as:

(a) overall signal strength at various points in the connection;

(b) the control of signal loss and the electrical stability of the connection;

(c) the limits on acceptable signal propagation time;

(d) the limit on acceptable noise disturbance;

(e) the control of sidetone and echo;

(f) the limits on acceptable signal distortion, crosstalk and interference.

Although historically prepared to cover voice performance, the recommendations also give information on the effects of the impairments on voiceband data. This aspect is becoming increasingly important with the widespread use of the public switched telephone network (PSTN) for accessing the Internet.

10.3.2 Transmission loss

Transmission loss is defined as the ratio of the input power to output power expressed in decibels[†] (dB). The loss across a voice connection consists of the

---

† Decibel (dB) is a convenient unit for measuring loss (or gain) of a network element or network, and is defined as loss $L = 10 \log_{10} P_1/P_2$ dB where P_1 is the input power to an element and P_2 the output power. Where a number of elements having gain or loss are connected in tandem, the overall gain or loss is the algebraic sum of gains and losses.

loss in translating sound into electrical inputs to the network, the loss across the network(s) and that incurred at the receiver. With no loss, the sound is too loud, but too much loss results in an inaudible signal. To facilitate network design, a system of 'loudness ratings' has been developed, namely:

(a) the 'send loudness rating' (SLR) is a measure of the signal loss expressed in decibels when the electrical signal amplitude is obtained from the original acoustic input; and

(b) the 'receive loudness rating' (RLR) of the receiver is a measure of the acoustic output obtained from the electrical signal input.

Determination of these loudness ratings for specific telephone instruments (and associated local lines) used to be carried out by subjective comparison between these instruments and standard reference instruments [4]. Today, computerised electro-acoustic measurement equipment is used to measure losses objectively and calculate loudness ratings.

The 'overall loudness rating' (OLR) is therefore the summation of send and receive loudness ratings plus the network loss, i.e.

$$OLR = SLR + \text{network loss} + RLR \text{ (decibels)}$$

This is illustrated in Figure 10.6, which shows the reference points at which SLR and RLR apply.

The end-to-end transmission loss must be acceptable to customers on an international connection and the ITU–T G series of recommendations

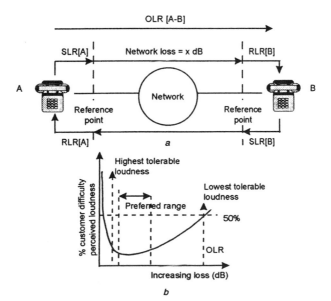

Figure 10.6   *Overall loudness rating*

      a Component losses of OLR
      b Variation with OLR of percentage of customers encountering
      difficulty in using connection

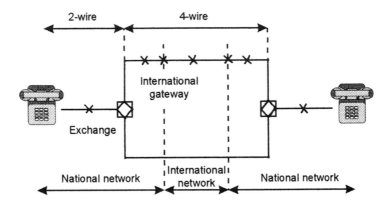

Figure 10.7   International transmission plan

specifies the transmission performance of the national and international components of the links, as shown in Figure 10.7.

10.3.3 Historical analogue transmission plan

For the old BT analogue network, the nominal maximum loss for a call routed between two local exchanges via two trunk exchanges in tandem was 19.5 dB. If the local access links had a maximum loss of 10 dB, an end-to-end loss of around 40 dB could be experienced, as shown in Figure 10.8. However, the number of calls experiencing such a loss would depend on the distribution of traffic and its routeing, and such calls could be quite rare.

For such a plan the nominal maximum OLR is:

$$OLR = SLR_{max} + 19.5 + RLR_{max}$$
$$OLR = 10 + 19.5 + 1 = 30.5 \text{ dB}$$

Rounded to 30 dB, this complies with ITU-T recommendations. However, 50% of people might subjectively encounter difficulty over a connection offering the maximum loss, as shown in Figure 10.8.

10.3.4 Digital transmission plan

A feature of digital transmission is that no loss is incurred over the digital path; however, connections with zero interexchange loss would result in calls being in a range approximately 6 dB louder than customers would prefer. Thus, it is necessary to design the digital switched network to a 2-wire-to-2-wire loss of about 6 dB on all classes of connection, assuming that line feeding and telephone instruments do not change.

With the Integrated Digital Network (IDN), the 2-wire local loop is converted to 4-wire at the local-exchange-line termination unit and then into

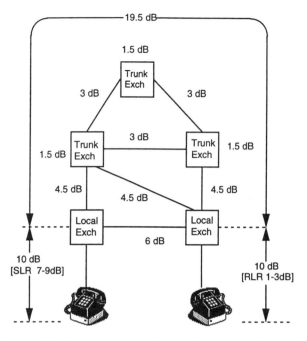

Figure 10.8 *Nominal limiting connections for analogue transmission plan*

the digital format with a net loss of 1 dB in the transmit direction; at the destination exchange the signal is reconverted to analogue with a further 6 dB loss giving an overall loss of 7 dB between reference points, as shown in Figure 10.9. Whereas a digital network gives superior transmission performance to analogue, the inherent loss of the analogue/digital conversion in digital exchanges can obviously degrade overall performance if they are introduced into an analogue switched network. Likewise, there are undesirable losses introduced when interworking between analogue and digital networks, which constrains traffic routeing options.

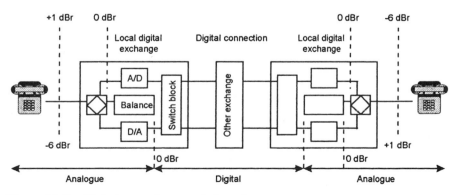

Figure 10.9 *Digital-network-transmission plan*

10.3.5 Stability

Insufficient loss around the 4-wire loop could result in positive feedback and the circuit oscillating in an uncontrolled manner; this is sometimes known as 'singing'. This condition could occur if the balance of the hybrid transformer which converts the 2-wire loop to 4-wire is disturbed, thus causing a low loss between the 4-wire connections. It is therefore necessary to have 'stability loss' around the 4-wire loop to prevent instability [5]. Stability loss is defined as the lowest loss between equirelative points at a 4-wire interface, from the receive to the send ports, measured at any single frequency in the band 0–4000 Hz. This measurement is applicable to all phases of the connection, e.g. connection establishment, information transfer and release.

A digital network has independent GO and RETURN paths, thus providing effective 4-wire working. The 6 dB loss provided on the single pair gives a stability margin of 12 dB, as illustrated in Figure 10.10.

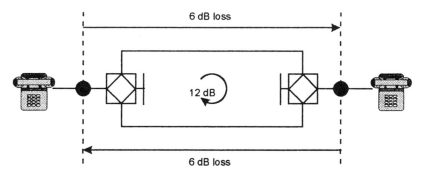

Figure 10.10   Stability margin

10.3.6 Echo

Echoes are caused by reflection of the speaker's voice back from the distant receiving end due to an imperfect line balance at the hybrid which causes part of the signal energy transmitted in one direction to return in the other. The signal reflected to the speaker's end is known as 'talker echo' and that reflected to the listener's end is called 'listener echo', this is illustrated in Figure 10.11.

The echo delay time is equal to the time taken for propagation over the connection and back again, and is therefore related to the line length, the delay introduced by switching systems, regeneration and signal processing. The obtrusive effect of echo on the customers increases with delay, ranging from a hollowness effect to large delay causing the speaker to interrupt his or her conversation and reducing the intelligibility of the received speech.

Echo can be reduced by sufficient loss in the echo path. This is known as 'echo loss' and is defined as 'the loss between equirelative level points at a 4-wire interface, from the receive to send ports measured as a weighted quantity in the frequency band 300 to 3400 Hz. The measurement is

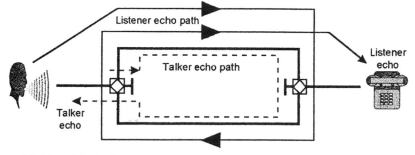

Figure 10.11   Echo paths

applicable during the information-transfer phase of the connection. ITU-T recommends [6] that the mean value of echo loss presented by a national network should not be less than $(15 + n)$ where n is the number of circuits in the national chain. Within a small country, like the UK, the relatively high loss of the old analogue network gave satisfactory echo performance. However, the reduced loss of digital networks, together with their increased delay, has required considerable care to avoid the expense of echo control. The 6 dB loss, together with the loss across the hybrid, is usually sufficient to suppress echo, but when digital networks are interconnected the additional delay could cause problems.

The ITU-T recommends that echo-control devices should be incorporated if the one-way delay of the echo path exceeds 25 ms [7]. Echo control is required on very long-distance calls, e.g. intercontinental calls, and this is carried out by echo suppressors or echo cancellors.

An echo suppressor is a voice-operated attenuator fitted to one path of the 4-wire circuit which is operated by speech signals on the other path, hence interrupting the echo path. Unfortunately, if the connection contains a number of circuits with echo suppressors connected in tandem, these can operate independently causing a lockout condition. Also, the switching time of the echo suppressors is generally too slow for data. A more satisfactory echo-control device, used in modern networks, is the echo canceller which cancels the echo by subtracting a synthesised replica of it, derived from the received signal.

10.3.7 Delay

Propagation delay is an inevitable consequence of transmitting signals over distance and one-way delay is defined as the time taken by a signal applied at the input of an equipment (or connection) to reach the output of that equipment (or connection). Excessive delay not only incurs the risk of echo but also impairs communication. For speech, it leads to confusion because of the response delay which can cause a speaker to repeat a sentence when the other party is responding. For data and signalling, where the protocol requires a rapid response, retransmission may occur with a consequent breakdown of communication.

The problem is most acute on transmission via satellites in geostationary orbit, where the one-way delay of 260 ms gives a pause between talker and response of about 0.5 seconds. The ITU-T recommends [8] that, except under exceptional circumstances, maximum one-way delay should not exceed 400 ms for an international connection. The one-way delay for connections within the UK is 23 ms. Connections that are routed within the UK network to, or from, an international gateway should not exceed a one way-delay of 12 ms.

Within the digital network, delay is increased due to the switching mechanism (which writes and reads from buffer stores), multiplexing and other signal processing. Table 10.2 illustrates typical delays [9]. This can be a problem when networks are interconnected, which often increases the number of exchanges and transmission systems traversed by an internetworked call.

Table 10.2   Typical delays

2 Mbit/s, pair cable	4.3 µs/km
140 Mbit/s coax cable	3.6 µs/km
140 Mbit/s, fibre	4.9µs/km
140 Mbit/s radio	3.3 µs/km
Digital mux/demux pair	1–2 µs (depending on hierarchical level)
Digital trunk exchange (sum of both directions, digital–digital)	<900 µs (mean value), 1500 µs (95% percentile)
Digital local exchange (sum of both directions, analogue–analogue)	3000 µs (mean value), 3900 µs (95% percentile)
Satellite geostationary orbit	260 ms

10.3.8 Error

An error is defined as a single-bit inconsistency between the transmitted and received signal. It is an important impairment in digital networks, particularly when they carry data services. To take account of the distribution as well as the number of errors, the two error-performance parameters, shown in Table 10.3 with performance objectives, are used; they apply to the 64 kbit/s level of the digital hierarchy. The international hypothetical reference connection used by the ITU-T [10] to specify error performance is shown in Figure 10.12.

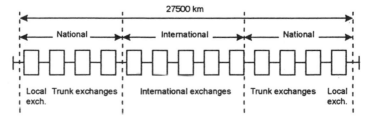

Figure 10.12   ITU-T hypothetical reference connection

Error-performance targets suffer because they cannot be extrapolated to higher levels in the digital-multiplexing hierarchy and they do not indicate the severity of bursts of errors.

Table 10.3   Error-performance objectives

Performance parameter	Definition	G821 recommendation	ITU-T International Reference Objective
Errored seconds	A time period of 1 s with 1 or more errors	<1.2% of 1 s intervals to have any errors	<8% of 1 s intervals
% error-free seconds		>98.8%	>92%
Severely errored seconds	A time period of 1 s with 65 or more errors	99.935% of 1 s periods with error ratio better than 10^{-3}	<0.2% of 1 s intervals with error ratio >10^{-3}

10.3.9 Slip

A synchronisation network is used to ensure that all digital exchange clocks are operating at the same average frequency. If not, the information rate of a signal received at an exchange would be different from the rate at which the exchange could process the information and retransmit it. This results in information being lost if the input rate is faster, or repeated if it is slower. This process is known as 'controlled slip'. If the unit of information which is deleted or repeated is a complete frame (256 bits), it is known as frame slip. Slips will occur at each switching unit in the IDN at a rate proportional to the frequency difference between the incoming bit streams and exchange clocks and will cause error degradation to services, particularly data services. There are two main forms of controlled slip namely:

(a) 'Octet slip', which is a single slip of 8 bits, representing the repetition or deletion of one sample of a 64 kbit/s pulse coded modulated (PCM) encoded signal.
(b) 'Frame slip', which constitutes a slip of one whole frame of a 2 Mbit/s primary-level digital signal.

Uncontrolled slip is the unforeseen event from a lapse in the timing process resulting in the deletion or repetition of bit(s). Since it is indeterminate, it would normally lead to the loss and subsequent recovery of frame alignment, with a potentially serious effect on the service carried.

The maximum theoretical slip rate for a single plesiochronous inter-exchange digital link, where the clocks of the digital exchanges have a long-term frequency accuracy better than one part in 10^{11} (controlled by the synchronisation network) is one controlled slip every 70 days. The ITU-T objectives for controlled slips in a hypothetical reference connection are contained in *Recommendation G.822* [11].

10.3.10 Jitter

Jitter [4] is defined as short-term variations of the significant instants of a digital signal from their ideal positions in time. Jitter arises from the way in which bit timing associated with a digital signal is extracted from the signal itself and from the 'justification' associated with higher-order systems. Low-frequency jitter can accumulate as the signal passes along a chain of systems. PCM-encoded speech is fairly tolerant of jitter but digital-encoded TV is much more sensitive to it.

Jitter amplitudes are defined in terms of the nominal difference in time, expressed as unit intervals, between consecutive significant instants of the digital signal. Jitter (and wander) levels can be maintained below maximum limits by controlling the 'transfer characteristics' of jitter (or wander) between equipment input and output ports.

10.3.11 Wander

Wander [4] is the long-term variation of the timing from the ideal due, for example, to temperature variations of the transmission media which, in turn, cause variations in propagation times. Typical variations, expressed in units of time, are 0.3–0.4 ns/km monthly variation and 0.6–0.8 ns/km annual variation for 140 Mbit/s systems on fibre-optic cable.

10.3.12 Sidetone

Telephones contain the equivalent of a 2-wire/4-wire hybrid to isolate the transmitter from the receiver. Leakage between transmitter and receiver is known as sidetone, and a small amount is necessary to make the telephone feel live to the user. An excessive level can make the talker speak too softly and also impair the received signal by excessive room noise transmitted as sidetone.

Sidetone is defined as the proportion of the talker's speech which is fed back to his/her ear and is measured in terms of a 'sidetone-masking rating' (STMR) which is the weighted loss, in decibels, between acoustic interfaces at the transmitter and receiver. The ITU-T recommends values of at least 12 dB [12].

10.3.13 Noise

The noise performance of a connection can best be described in terms of three separate parameters, namely:

(a) *Impulsive noise:* This occurs when a noise burst of a defined threshold power and duration is detected at the output of equipment. It is generally experienced as the 'clicks and bangs' one hears during a telephone conversation. It is usually generated by the older switching and signalling systems, together with poor joints in the local loop.

(b) Psophometrically weighted noise: This is a relatively low level of background noise or 'hiss' which is present throughout the information transfer phase. It is measured as the total noise in the speech band of 300–3400 Hz with a filter which has the weighting characteristic similar to the gain/frequency response of a telephone receiver and the ear. The noise is measured in terms of power, and typical ITU-T objectives for analogue FDM systems are 3 pW0p/km (pW = picowatt and 0 indicates that the value is measured relative to the 0 dB reference point and p = psophometric noise) for line circuits and 200 pW0p for channel modulators.

(c) Single frequency noise: This is defined as the noise power at a discrete frequency. This type of noise in the speech band can be very annoying and needs to be at least 10 dB lower than than the psophometrically weighted noise for it to be masked by the background noise.

10.3.14 Quantisation distortion (quantisation noise)

The process of converting an analogue signal into a PCM digital signal involves a process of sampling the analogue waveform at regular time intervals and measuring its amplitude, then translating these amplitudes into a coded binary digital signal.

Since the quantisation process turns a continuous signal into discrete voltage steps (see Figure 10.13), it is not possible to reconstruct the original signal except as an approximate representation. Hence nonlinear quantisation distortion is introduced which, on voice communication, appears as noise; thus it is often referred to as 'quantisation noise'. The smaller the quantising levels the less will be the distortion, but this will require more digits in the code words representing the levels, which increases the bandwidth required to transmit them.

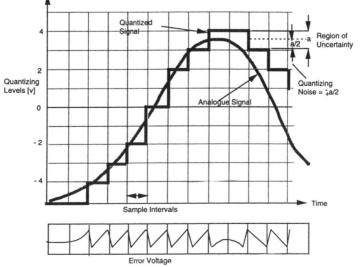

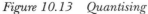

Figure 10.13 Quantising

This type of distortion is expressed in terms of 'quantisation-distortion units' (QDU). One PCM system using 8-bit coding with 256 quantising levels is considered to produce 1 QDU which is equal to a signal-to-distortion ratio of about 36 dB. When *n* such PCM systems are cascaded, then *n* QDUs are amassed. One 7-bit PCM system, with fewer quantisation levels, is considered to produce 3 QDUs which equates to a 6 dB degradation of the signal-to-distortion ratio. For an international link, ITU-T recommends a maximum of 14 QDU [13], four for the international link and five for each inland connection. For integrated digital networks, connections will have 1 QDU because multiplexing and demultiplexing only take place at the originating and destination ends of a call.

10.3.15 Loss distortion with frequency

This is defined as the ratio of output voltage at a reference frequency divided by its value at any frequency in a specified band, with the input-signal level constant, expressed in decibels. It is often loosely referred to as frequency response and observed as a variation of transmission loss with frequency. ITU-T recommendations for frequency division multiplex (FDM) speech channels suggest a maximum differential attenuation of 9 dB, using appropriate 'equalisation' to ensure that the frequency response of the circuit is within the 'mask' illustrated in Figure 10.14.

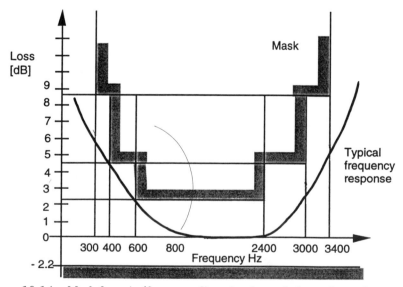

Figure 10.14   Mask for gain/frequency distortion in a telephone channel

10.3.16 Group-delay distortion

When the time taken to traverse a system is a function of the frequency, signals at different frequencies arrive at different times, and the variation in this delay is called group-delay distortion [14]. It is defined as the envelope delay measured at the output at a given frequency, compared with the minimum measured envelope delay at any frequency in a specified frequency band. The effect is also a differential shift in phase over the frequency band, and this is inevitable when a circuit contains transformers and coupling capacitors.

Subjectively, the impact on speech transmission is negligible since the ear is insensitive to differences in phase. However, the effect on a pulse shape is to spread it in time; hence if the signal waveshape is to be preserved (e.g. for data and TV signals) then the distortion must be minimised.

10.3.17 Crosstalk

Crosstalk may be defined as the presence of unwanted signals coupled from a source other than the connection under consideration. For speech, even if low-level crosstalk is unintelligible, it can give the user an impression of lack of secrecy and it is therefore an undesirable reduction in QoS.

Crosstalk occurs in multipair cables in the local loop by capacitive coupling between the pairs. It can also be caused by nonlinearity in FDM systems from which intermodulation products of frequencies in one channel may fall within the frequency bands of other channel(s). Crosstalk is primarily affected by:

(a) the level of signal: if high there is more likely to be crosstalk into another channel;

(b) the frequency of interfering signal: there is more likely to be crosstalk as it increases;

(c) the balance of the circuits about earth: if wires in a cable have equal capacitance to earth and to each other, the crosstalk between pairs will be zero. Unbalance increases the risk of crosstalk as does balance to earth of the impedance terminating the line;

(d) the quality of the screening between circuits.

When crosstalk is transmitted over the disturbed channel in the same direction as its own signal, it is known as 'far end crosstalk' (FEXT); when it is transmitted in the opposite direction to its own signal it is known as 'near end crosstalk' (NEXT), as shown in Figure 10.15. However, crosstalk can usually be detected at both ends and NEXT and FEXT refer to the near- and far-end measurements. If the power of the crosstalk signal, measured at the near or far end of the disturbed channel, is P_1 and the power of the disturbed channel's own signal is P_2, then the

$$\text{crosstalk ratio} = 10 \log 10 \, P_2/P_1 \text{ decibels}$$

The crosstalk ratio is usually measured when test signals of the same level are applied to both the disturbing and disturbed channels.The risk of crosstalk being audible is negligible if the crosstalk ratio exceeds 65 dB [14].

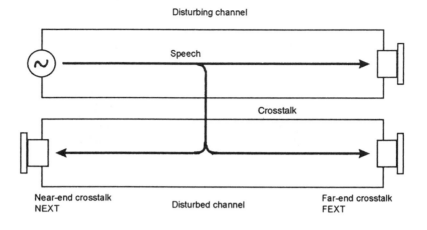

Figure 10.15   Crosstalk

10.4 Availability

This parameter describes how much of the time a system or network will be available and operating, measured as the ratio of uptime to total time, as a percentage. From a customer perspective, unavailability is a more meaningful measure since it represents the time during which a failure is experienced, generally expressed as annual downtime. Some customers, particularly data users, are more concerned about the frequency of service disruption rather than to the total downtime. Overall unavailability is heavily influenced by the reliability of equipment which is generally measured in terms of 'mean time between failures' (MTBF) measured in hours or years, 'mean time to repair/replace/restore' (MTTR) or 'mean down time' (MDT). 'Mean time to failure' (MTTF) describes the uptime. MTTR and MDT are usually measured in hours. For components and small assemblies, whose mean lives are very long, the term 'failure rate' is used, usually measured as 'failure in 10^{-9} item hours' (FIT), Figure 10.16 illustrates the relationship between some of these parameters.

In theory, the MTTF of an item of equipment can be predicted from the expected reliability of its components [15]. The calculations become difficult due to design measures taken to improve reliability, such as redundancy. Such predictions are carried out by manufacturers to provide evidence of compliance with required MTBF specifications, but they are unreliable and usually do not match achieved MTBFs. Achieved MTBFs should be observed statistically over a sufficiently large sample. Typical values are 100 km fibre system = 3 years, 2–8 multiplex equipment = 30 years and exchanges = 50 years [16]. MTTRs can often be inferred from performance targets, e.g. percentage of faults cleared within x hours. They can be customer specific,

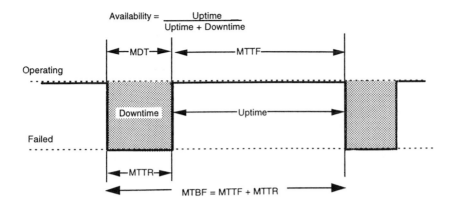

Figure 10.16   Availability parameters

depending on the level of service care contracted, e.g. a private-circuit 'reduced charges scheme' which guarantees that customers are rebated up to 100% of rental charges if six faults during a year are not cleared within 5 hours.

The impact of unavailability on customers may be expressed as:

(a) Unacceptability: The user deems the service effectively unavailable due to the poor performance, e.g. bit-error rate, that is encountered. However, this is user specific, since error on speech is not serious but can have a major impact on a data connection.

(b) Partial inaccessibility: The customer is unable to access parts of the network due to, say, congestion or isolation of a local exchange or isolation from another part of the network.

(c) Total inaccessibility: This occurs when the customer is completely isolated from the network. This could impact on one customer for a short break, e.g. ITU-T defines a complete failure on a 64 kbit/s connection as when the error ratio exceeds 1×10^{-3} in each second for more than 10 consecutive seconds. More serious examples impact on groups of customers, the larger the group the more serious the isolation. Since business and society are becoming ever more dependent on telecommunications, the impact of service outages becomes more severe. In particular, total network failures, sometimes referred to as 'brownouts', can have a major economic impact on a country. Moreover, the more technologically sophisticated the network becomes the more it is susceptible to such occurrences. Hence, it is essential that future network designs are resilient and self healing. These major failures are often classed as 'major service failures' and the mean time between them is known as MTBmsF. They are defined as a failure which results in either exchange isolation (i.e. the loss of all incoming or outgoing service or standard tone(s)

for 30s or more), or restricted service from the loss of 30% or more circuits in a route, loss of access to 50% of exchange traffic-carrying capacity or loss of service to 50% or 500 customers. Aspects related to major network failures are discussed in more detail in Chapter 11.

10.5 Call-processing performance

Performance parameters can be classified in terms of their impact on the processing of calls [9], i.e. access-to-network (or call-establishment) phase, or disengagement phase, information-transfer phase and availability. Most parameters described so far have been mainly concerned with the information-transfer phase and availability. Those effecting the call-processing phases are:

(*a*) *Delay to dial tone:* the interval between the instant a valid seizure signal is received by the network (local exchange) and the instant a dial tone is returned by the network.

(*b*) *Removal of dial tone:* the ability of the network to remove the dial tone on receipt of the first dialled digit. Failure to remove the dial tone results in an unavailability condition.

(*c*) *Connection-establishment delay:* the time from the instant the last digit of a valid destination address is received by the network (originating local exchange) to the instant a ring tone or number-engaged tone is returned by the network. It could also refer to the delay before tones/announcements indicating other network conditions.

(*d*) *Connection-establishment failure due to network congestion:* following the input of valid address information by the customer, connection to the required destination is not established due to network congestion. This aspect is explored in more depth in Section 10.6.

(*e*) *Connection-establishment failure due to no tone:* following the receipt of valid address information, connection to the required destination is not established and no tones or announcements are returned from the network.

(*f*) *Misrouteing of calls:* the network connects the call to a destination other than that requested.

(*g*) *Incorrect network tones or announcements:* the network returns a tone/announcement which is incorrect for the circumstances applicable.

(*h*) *Premature release of established connections:* the network releases an established connection when the conditions for release have not been met.

(*i*) *Delay to connection release:* the time from the instant the connection release request is received by the network to the instant all circuits in the connection are released and are ready to be used in establishing other connections.

(j) Connection release failure: the network does not release an established connection after the conditions for release have been met.

10.6 Network congestion

One of the most common causes of poor QoS is traffic congestion. The traffic demand on a network fluctuates in both the short and long term (i.e. over 24 hours). The calls in progress tend to peak during the mid-morning and early afternoon due to business use. A smaller peak usually occurs during the evening for social calls, and very few calls occur during the night. The sizes of the telephone exchanges and traffic routes which interconnect them are based on the predicted calls and their duration (sometimes called the holding time) during the busiest hour, i.e. the busy hour. However, the time of the busy hour can differ between exchanges according to the telephone habits of the community of customers served by them, e.g. whether predominantly business or residential, and the distribution of their calls.

The number of calls in progress fluctuates randomly as individual calls commence and end. The traffic is the mean number of calls in progress and, although this is a dimensionless quantity, the unit of traffic intensity is called the erlang. The traffic carried by a system or group of circuits is given by:

$$A = \frac{Ch}{T}$$

where A = traffic in erlangs, C = average number of calls arriving in time T (usually the busy hour) and h = average call duration.

It is not economical to to provide sufficient exchange and route capacity to cater for the greatest possible traffic demand that could arise. Hence, a proportion of call attempts will be rejected during the busy hour because there is insufficient equipment capacity, i.e. congestion will occur. This is known as the grade of service (GOS).

In a public switched telephone network (PSTN), attempts to make calls over congested exchanges or routes will not be successful and this is known as a lost-call system. In a message-switched system, e.g. a packet-switched data network, calls which arrive during congestion wait in a queue until capacity becomes available, i.e. calls are delayed but not lost. Such systems are known as delay or queuing systems.

When congestion occurs in a lost-call system [17] the traffic carried is less than that offered, hence:

traffic carried = traffic offered − traffic lost

The grade of service B is defined as:

$$B = \frac{\text{number of calls lost}}{\text{number of calls offered}}$$

hence, also

$$B \quad = \quad \frac{\text{traffic lost}}{\text{traffic offered}}$$

$\quad\quad\quad$ = $\quad$ proportion of the time for which congestion exists

$\quad\quad\quad$ = $\quad$ probability of congestion

$\quad\quad\quad$ = $\quad$ probability that a call will be lost due to congestion

Thus if A erlangs of traffic is offered to a route having a grade of service B

$$\text{traffic lost} \quad = \quad AB \text{ erlangs}$$
$$\text{traffic carried} \quad = \quad A(1 - B) \text{ erlangs}$$

Modern telecommunications networks are usually resilient against small unexpected surges of traffic because of the traffic-routing strategies employed. For example, automatic alternative routing (AAR) is used where calls encountering congestion on their primary route will overflow to an alternative route which may have free capacity; also, dynamic routing is an adaptive method which finds a free path for a call, from source to destination. However, such methods cannot cope with large unforseen increases in call-holding time or major surges in traffic.

Since the network is dimensioned to meet the forecast traffic demand, it follows that any unforecast increase in calls or their duration during busy periods is likely to cause congestion. In particular, an unforecast increase in average call-holding time, due to the introduction of new services such as access to the Internet, can drive the network into congestion, e.g. a single one hour Internet session is equivalent to 20 three minute telephone calls. This problem is likely to be exacerbated in the future broadband network where, for example, a single one hour 2 Mbit/s video-on-demand session is equivalent to 600 three minute telephone calls at 64 kbit/s.

A feature of the current telecommunications environment is the unpredictable and volatile nature of traffic where, for example, large surges of traffic can flow over the network causing severe congestion. Such events are often media driven, caused by television or radio phone-ins which result in traffic overloads focused on specific telephone numbers. Such overloads also block calls to other numbers on the same exchange and calls to other destinations sharing the same traffic routes. Such situations are exacerbated by customer-generated repeat attempts which can be higher than 90% [18]. The impact of such events can be minimised by network-traffic-management techniques such as call gapping. This blocks traffic destined for a congested destination as near to its originated source as possible to allow traffic to non-congested destinations a better chance of successful completion, e.g. 1 in n calls or 1 call every T seconds is allowed on to the network.

Congestion problems can also effect ITU-T No 7 signalling networks due to the high volumes of initial address messages generated by the high calling rates associated with traffic congestion, as can the increasing number of intelligent network services requiring multiple noncircuit-related signalling messages which can cause large volumes of signalling traffic.

Congestion in cellular-mobile-radio systems is a particularly complex problem. In addition to the problems of sizing of cells and allocation of radio frequencies to meet the needs of a random population of mobile users [19], there is the need to deal with call handover from one cell to another which is a form of overflow traffic where blocking is unacceptable.

10.7 Network-interconnect apportionment

When a connection spans more than one network, end-to-end performance targets must be apportioned over the interconnected networks. Figure 10.17 illustrates the increasing complexity of performance engineering when dealing with interconnected networks and the difficulties of apportioning a number of performance parameters over different networks to give a satisfactory end-to-end QoS.

The UK regulator (OFTEL) originally derived a simple performance model to assist in determining the performance requirements at points of interconnect between the BT and MCL networks. It divided the routeing of a call into three basic areas, namely 'collection system', 'national transport system' and 'delivery system'. The model is illustrated in Figure 10.18 together with similar models for international calls.

Much work needs to be carried out to determine performance-parameter values for network interconnect in the UK environment of multiple operator networks where delay, quantising distortion and echo are critical parameters. The simple model illustrated in Figure 10.18 is difficult to apply with multiple network operators, where the overall allocation may need to be apportioned between four or five operators.

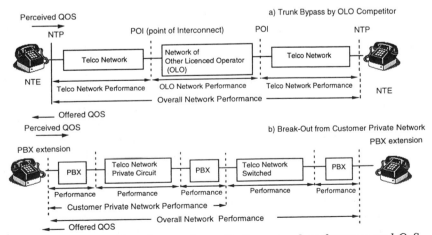

Figure 10.17   Examples of connections affecting network performance and QoS

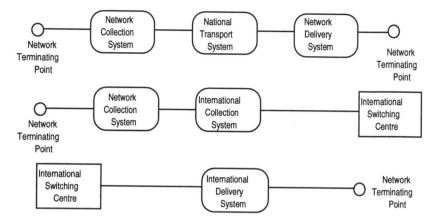

Figure 10.18   OFTEL internetwork-apportionment model

10.8   Service-related performance

Telecommunications services are currently growing in their numbers and variety and many require completely different performance. For example, data require good error performance but can tolerate reasonable delay, whereas speech requires the opposite. New services with characteristics which cannot easily be catered for in the PSTN have traditionally been dealt with by establishing separate networks, e.g. telex, data etc. In some cases internetworking gateways have been set up to provide a common customer access, generally through the PSTN. More recently, a multiservice network is being introduced in the form of the ISDN. However, this circuit-switched network is limited in bit rate to 64 kbit/s (i.e. narrow band, N-ISDN) and restricted in its capability to handle multibitrate calls or bursty information. It is now accepted that there is a potential demand for a variety of broadband services at different bit rates and traffic characteristics which cannot be carried on the N-ISDN. This demonstrates a need for a multiservice broadband ISDN (B-ISDN).

10.9 B-ISDN performance standards

The ITU-T has assumed that asynchronous-transfer-mode (ATM) technology will be the basis for the future B-ISDN and is formulating performance standards to meet the expected range of services. The ITU-T [20] has classified services as interactive or distribution, as illustrated in Figure 10.19.

ATM is a packet-based communication system (although the packets are usually referred to as 'cells'). This means that there is the potential within the network for information loss and delay. The network-performance parameters of an ATM-based B-ISDN network are specified by the ITU-T in Recommendation I.350 [21]. The main parameters are:

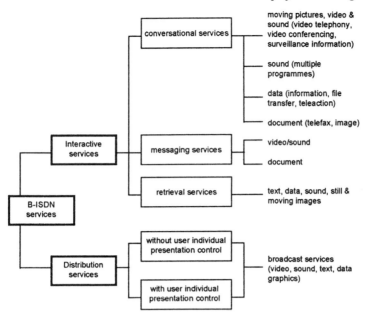

Figure 10.19   Classification of B-ISDN services

Call-control parameters

(a) *Connection-set-up delay:* the time interval between the call-set-up message transfer and the call-set-up acknowledge-message-transfer event without the called-user response time;

(b) *Connection-release delay:* the time interval between the call-release message-transfer event and the call-release acknowledge-message-transfer event;

(c) *Connection-acceptance probability:* the proportion of accepted calls experienced by a customer over a long period of time.

Information-transfer parameters

(d) *Cell-error ratio (CER):* the ratio of total errored cells to the total of successfully transferred cells plus errored cells in a population of interest;

(e) *Cell-loss ratio (CLR):* the ratio of lost cells to the total transmitted cells. The loss of cells can occur due to errors in the header or due to buffer overflow in an ATM switch. This parameter can also be expressed as cell-loss probability (CLP);

(f) *Cell-misinsertion rate (CMR):* the total number of misinserted cells observed during a specified time interval divided by the duration of the time interval;

(g) *Cell-transfer delay (CTD):* the time between the occurrence of two corresponding cell-transfer events;

(h) *Cell-delay variation (CDV):* the difference between the values of the transit delay of cells belonging to the same broadband communications channel during a predefined period of time;

(i) Severely errored cell-block ratio (SECBR): the ratio of total severely errored cell blocks to total cell blocks in a population of interest.

These performance parameters can be mapped across to the general QoS characteristics as shown in Table 10.4

For voice services, delay is important and can be objectionable if echo is present. As delay increases to about 20–30 ms, conversation can sound reverberent and some echo can begin to be perceived. Delays may be caused by cell assembly, transmission and buffering. Cell assembly is the time taken to fill a 48-octet cell with speech samples, i.e. for speech encoded at 64 kbit/s it is 8 bits per sample/64 kbit/s $\times$ 48 = 6 ms.

Cell loss can arise from addressing errors or discarding cells under network overload. A lost cell is replaced by a cell of 'all 1s' which can cause an audible click and be objectionable at high cell-loss ratios.

For video services, cell-loss effects are important because much of the redundant information in the original video signal is discarded by the compression process. Delay is not important for broadcast video services, but it can impede the exchange of information when conferencing. Delay jitter can, however, cause loss of video-frame synchronisation and consequential degredation of picture quality. Some data services, such as file transfer, require good data integrity, i.e low errors and cell loss.

Table 10.4 Mapping of QoS and performance

QoS characteristics	ATM performance parameters
Time-related characteristics (time delay)	Cell inter-arrival time CDV (delay and jitter) CTD
Capacity-related characteristics (communication throughput) Guarantee level	Cell-transfer capability Mean cell rate Best-effort class (unspecified QoS class) Statistical/compulsory (specified QoS class)
Delay	CTD Mean CTD
Channel availability (the proportion of time for which the communication channel is available)	Priority level (loss priority level) Cell-loss probability
Accuracy (error-related parameters)	Cell-sequence integrity CLR CER CMR SECBR Cell-insertion rate Bit-error probability

Clearly, the ATM-based B-ISDN will support a wide range of services with diverse QoS and, hence, network-performance requirements. Some services will require stringent performance criteria, but others can operate with more relaxed performance criteria and still deliver acceptable QoS. Giving all services the same, most stringent, performance criteria will require more network capacity than is necessary; hence, the ITU-T [20] has defined a series of QoS classes, each with common performance requirements. This will allow the network to provide the appropriate performance for each service delivered, with consequential increase efficiency. These classes are:

Class 1 (stringent): specified for real-time rate-oriented services, examples of which are shown in Table 10.5.

Class 2 (tolerant) and Class 3 (bilevel): no CDVs are specified but there are commitments to CLRs. These classes are for nonreal-time, unit-oriented applications which transport discrete units of information, i.e. messaging and retrieval services where delay is not an issue.

Class U (unbounded): where performance parameters are not specified and can be worse than the other classes. Typically, this class is used for data applications where delay is unimportant and errors can be compensated for by retransmission, which, in turn causes delay.

Table 10.5   Examples of class 1 services

Application	Bit rate	CDV	CTD	CLR
Telephone[†]	64 kbit/s	20 ms	400 ms	10^{-3}
Videophone[†]	64 kbit/s-2 Mbit/s	20 ms	200 ms	8×10^{-6}
Videophone[‡]	2 Mbit/s	20 ms	200 ms	8×10^{-6}
Videoconference[‡]	5 Mbit/s	20 ms	200 ms	8×10^{-6}
TV distribution[†]	20–50 Mbit/s	20 ms	no limit	8×10^{-7}

[†] *Constant-bit-rate (CBR) services*
[‡] *Variable-bit-rate (VBR) services; bit rates shown are the average values of the variable bit rates*

The European Commission has sponsored many research projects in ATM networking (mostly under the RACE programme, see Chapter 3), and some have been concerned with QoS. The report of RACE 510 considers four major classes of user:

• large corporate;
• small corporate;
• domestic; and
• people with special needs.

For all of these, the effect of the varying network performance (NP) will be different. The user perceived quality is determined by NP, but not in so direct a fashion as seeing x cells lost in a connection!

Also important is the effect on the two different classes of service: real time, and nonreal time. In the former, all cell delay is either insignificant, or is effectively cell loss; in the latter, all cell loss translates to cell delay, as lost cells are simply retransmitted. So, in line with RD 510 [22], it is assumed that:

(i)   real time means those connections for which delay translates to loss, i.e. there will be no retransmissions at the transport layer (layer 4);

(ii)   nonreal time means those connections for which loss translates into delay, i.e. there will always be a retransmission at the transport layer.

The network performance offered by the network operators must be carefully 'mapped' in order to find the users' perceived GOS. This can be illustrated by considering, initially, voice connections, for which we can consider all those cells whose delay is such as to render them effectively lost. Assume that the CLP in ATM networks is $\sim 10^{-9}$. This is an example of NP, which we can translate into GOS.

(a) Mapping voice
Assume:

- 3 min mean holding time
- 64 kbit/ connection

Cell rate   = 64 kbit/s/(8×48) bits/cell = 167 cells/s
    Then the mean number of cells per connection is

$167 \times 3 \times 60$ $\simeq 30\ 000$

And therefore the probability that the connection is affected by cells(s) lost is

$1 - (1 - 10^{-9})^{\,30\ 000} = 3 \times 10^{-5}$

That is to say three calls in every 100 000 (on average) will be affected by >0 cell(s) lost. A figure which is more comprehensible to a user is that 99 997 calls in every 100 000 should be error free.

    This can be further translated. If a domestic user, on average, makes one call per day and receives one call per day, then the user will be affected, by cells(s) lost during a connection, every

$33\ 333/(365 \times 2) = 45$ years on average

A corporate user may make and receive 500 calls a day, so here the figure will be cell(s) lost during a connection every

$33\ 333/500$ = 2 months (approx.) on average

(b) Mapping video
A similar first order calculation can be made for a real-time video link. Assuming:

- 1 Mbit/s
- average length of connection = 1 hour

Similar calculations as for voice give:

    average number of cells per connection   = 10 million

    probability that connection is unaffected by cell(s) lost = 0.99

So, on average, 1 in every 100 will be affected. At a rate of one connection per week, a user will average 99 weeks before a connection is affected.

(c) Mapping of a nonreal-time traffic

If it is assumed that all loss now produces delay, then in the event of an error, packet retransmission (of the packet) is required, and one packet can be assumed to be 64 cells. Assume a total of 100 Mbyte of data to be transmitted.

Example 1

100 Mbytes of data	= 32 552 packets
	= 2 083 328 cells
CLP	$= 10^{-9}$
probability of a packet losing a cell	$= 1 - (1 - 10^{-9})^{64}$
	$= 6.4 \times 10^{-8}$

Assume that the distribution of lost cells within any packet is binomial; then:

average number of packets lost in 32 552 = NP
$$= 6.4 \times 10^{-8} \times 32\ 552$$
$$= 2 \times 10^{-3}$$

Which is so nearly zero that we can conclude that the users will (for CLP=10^{-9}) see virtually no extra delay in their data service due to data retransmission.

Example 2

If we consider a much higher CLP, say CLP $= 10^{-3}$, then we can recalculate the effect of retransmissions.

probability of packet losing a cell	$= 1 - (1 - 10^{-3})^{64}$
	$= 0.062$
average number of packets lost in 32 552	= 2018

But when these are retransmitted some of these will also be errored, so:

average number of packets lost in 2018 = 125

And some of these will be errored:

average number of packets lost in 125	= 7
average number of packets lost in 7	= 0.48

Thus in total there will be (on average) only three retransmissions in successfully moving 100 Mbyte of data across a network with CLP = 10^{-3}. So it may be provisionally concluded that even 'high' CLP's may not badly affect the user perceived GOS of nonreal-time data services.

These calculations have required certain simplifying assumptions, and these include the assumption that the *cell-loss events occur at random*. Experience, and some analytically modelling, have shown that this will probably not be the case. Cell-loss *events* will tend to be clustered and occur with a probability which is less than the unconditional CLP. The effect of this will probably be to make connections affected by cell loss(es) rarer, but ensure that they are affected more severely when loss does occur.

10.10 Conclusions

This chapter has briefly described the primary network-performance parameters which have a significant impact on QoS and, hence, in a competitive environment, on market share. The increasing change to networks together with the introduction of new services makes performance engineering a complex task, and it is necessary to issue detailed guidance to all concerned with the planning and operation of the network and services it supports.

10.11 References

1   ITU-T Recommendation I.350: 'General aspects of QoS and network performance in digital networks including ISDNs' (International Telecommunications Union, 1993)
2   ITU-T Recommendation E.810: 'Framework of the Recommendations on the serveability performance and service integrity for telecommunication services' (International Telecommunications Union, 1992)
3   ITU-T Recommendation M.3200 series: 'Telecommunications management network' (International Telecommunications Union, 1992)
4   McLINTOCK R. W.: 'Transmission performance' *in* FLOOD, J. E., and COCHRANE, P. (Eds.): 'Transmission systems' (Peter Peregrinus, 1991)
5   FLOOD, J. E.: 'Telecommunications switching, traffic and networks' (Prentice Hall, 1995), chap. 2
6   ITU-T Recommendation G.122: 'Influence of national systems on stability, talker echo in international connections' (International Telecommunications Union, 1993)
7   ITU-T Recommendation G.131: 'Control of talker echo' (International Telecommunications Union, 1996)
8   ITU-T Recommendation G.114: 'One-way transmission time' (International Telecommunications Union, 1996)
9   COOK, G. J.: 'Network performance' *in* FLOOD, J. E. (Ed.): 'Telecommunications networks' (IEE, 1997)
10 ITU-T Recommendation G.821: 'Error perfromance of an international digital connection operating at a bit rate below the primary rate and forming part of an integrated services digital network' (International Telecommunications Union, 1996)
11 ITU-T Recommendation G.822: 'Controlled slip rate objectives on an international digital connection' (International Telecommunications Union, 1988)
12 ITU-T Recommendation G121: 'Loudness ratings of national systems' (International Telecommunications Union, 1993)
13 ITU-T Recommendation G.113: 'Transmission impairments' (International Telecommunications Union, 1996)
14 FLOOD, J. E.: 'Transmission principles' *in* FLOOD, J. E., and COCHRANE P. (Eds): 'Transmission systems' (Peter Peregrinus, 1991)
15 GHANBARI, M., HUGHES C. J., SINCLAIR M. C., and EADE J. P.: 'Introduction to reliability' *in* 'Principles of performance engineering for telecommunications and information systems' (IEE, 1997, chap 9)
16 FLOOD J. E.: 'Telecommunications switching, traffic and networks' (Prentice Hall, 1995), chap. 2
17 FLOOD, J. E.: 'Telecommunications switching, traffic and networks', (Prentice Hall, 1991) chap. 4
18 SONGHURST, D. J.: 'Teletraffic engineering' *in* FLOOD, J. E. (Ed): 'Telecommunications networks' (IEE, 1997)

19 MACFAYEN, N. W., and EVERITT, D. E.: 'Teletraffic problems in cellular mobile radio systems'. *11th International Teletraffic Congress*, 1985, paper 2.4B-1
20 ITU-T Recommendation I.211: 'B-ISDN service aspects' (International Telecommunications Union, 1993)
21 ITU-T Recommendation I.356: 'B-ISDN ATM layer transfer performance' (International Telecommunications Union, 1993)
22 RACE draft specification RD510: 'General aspects of QoS and network performance in IBC', 1993

Chapter 11
Network integrity

11.1 Introduction

Network integrity can be broadly defined as 'the ability of a network to retain its specified attributes in terms of performance and functionality'. Clearly, isolated incidents which impact only on a small number of customers should not be classed as a breach of network integrity. It is therefore generally accepted that the definition should focus on major outages or performance degradation affecting large numbers of customers. The exception is where loss of an important, say, intelligent network (IN) service is experienced where a relatively small number of major business customers subscribe to it. The US Federal Communications Commission (FCC) defines an outage as 'a significant degradation in the ability of a customer to establish and maintain a channel of communications as a result of failure in a carrier's network'.

The widespread use of processor-controlled technology in the networks with high-capability common-channel signalling and 'intelligent' platforms creates an increasingly unstable equilibrium where comparatively minor perturbations can cause severe network outages. The vulnerability of these new technologies is exemplified by the severity of the signalling-related failures (sometimes known as 'brown-outs') which affected several network carriers in the USA during 1990/91. For example, in January 1990, over half of the AT&T network (in the USA) failed for a period of around 9 hours due to a software problem, causing severe disruption to long-distance services [1]. The outage was due to a single-bit 'soft glitch' cascading through the SS7 network (ITU-T Signalling System 7) which was subsequently found to be a single AND instead of an OR condition in several millions of lines of software code [2]. In June and July 1996, there were major outages which affected millions of customers of Bell Atlantic and Pacific Bell (USA) caused by a faulty software upgrade to SS7 Signal Transfer Points (STPs) [3].

These incidents prompted the US Congress to criticise the FCC severely for not being proactive in protecting the communications infrastructure [1]. Congress concluded that:

(a) the public switched networks are increasingly vulnerable to failure: and the consequences for consumers and businesses and for human safety are devastating;

(b) the problem of network reliability will become increasingly acute as the telecommunications market becomes more competitive; and

(c) no Federal agency or industry organisation is taking the steps necessary to ensure the reliability of the US telecommunications network.

Subsequently, the FCC established the Network Reliability Council and gave the remit to the Alliance for Telecommunications Industry Solutions (ATIS) to produce monitoring and measurement proposals [4]. The FCC has imposed a reporting regime which requires that it be notified of all outages of at least 30 minutes and effecting 30 000 or more customers, plus outages affecting emergency services and key facilities.

The potential problems due to network complexity are exacerbated when networks are interconnected. This is becoming a major regulatory issue, since such policy is aimed at allowing unrestricted interconnection between the networks of established network operators and competing operators and service providers. An example is the European Union Open Network Provision (ONP) policy. The European Commission has recognised the risks of this policy to network integrity and has commissioned work to determine how the risk can be minimised [4]. In the UK, the regulator (OFTEL) has established the Network Interfaces Co-ordination Committee (NICC) as the consultative forum to advise the regulator in relation to optional and essential interfaces considered important in the progressive deregulation of telecommunications systems and services.

11.2 Network-integrity definition

Since the definition of network integrity is 'the ability of a network to *retain* its specified attributes in terms of performance and functionality', it considers integrity in terms of robustness, invulnerability and incorruptibility. In this sense, the degree of integrity of a system is inversely related to the risk of failure, i.e. the higher the degree of integrity, the easier it is for the system to maintain its attributes, and hence the less likely it is that unexpected perturbations can result in failure. The degree of integrity can be divided in to a collection of states of integrity, illustrated in Figure 11.1 [5]. At one end, there is 100% integrity, which means that the system is absolutely robust and will not be affected by unexpected problems. At the opposite end, there is 0% integrity, meaning that any perturbation will cause failure. Between those two limits, the network can be at different states of integrity.

The different bands between 0% and 100% integrity should be represented as a continuous gradient. However, this is not feasible in practice. Owing to the great complexity of the problem, the parameters introduced to measure integrity must take discrete values. Hence there is a need to introduce real boundaries between bands, with different values of integrity. The separation between these bands will not always be strict and there may be overlaps.

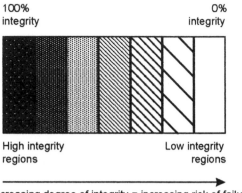

Figure 11.1   Integrity scale

Breach of network integrity and failure are related concepts, but network integrity goes beyond failure identification. Loss of integrity means moving from one band to another band of lower integrity, but this does not necessarily imply failure. Breach of integrity would imply failure only in one of the following cases:

(a) if the system was initially in a band of very low integrity, because any minor perturbation would put it in a situation of failure; hence, it is important to operate in a high-integrity band;

(b) if the loss of integrity is very high, causing the system to move from a band of acceptable integrity to the region near 0%. In some cases, loss of integrity cannot be avoided, e.g. there may always be unforeseen problems which, depending on the state of the system, might cause a chain reaction, leading to serious failure. However, it is desirable to minimise such instances, either by identifying possible sources of failure and designing them out, or by monitoring the relevant parameters to take avoiding action.

11.3 Network control

Modern networks are software controlled and the services they provide are established by the interaction of application programs in appropriate network elements via messaging over the signalling network, as shown in Figure 11.2. Interactions also occur between support systems and embedded network intelligence for the purpose of service and network management. In the future, it is increasingly likely that information technology (IT) applications in customer-terminal equipment for multimedia services will require heavy interaction with network control. Should invalid messages be exchanged between application programs, then misoperation can disrupt the normal operation of the network. In extreme cases, such a malfunction can cause invalid messages to propagate through the network, closing it down. Such problems can be caused by hardware faults, software 'bugs', operational errors, deliberate sabotage or incompatibilities in signalling and/or software. When constituent parts of the networks of different network operators and service providers are interconnected the probability of such perturbations occurring rises and causes threats to network integrity. Moreover, the behaviour of such interconnected software resources is nonlinear and possibly chaotic, and hence almost impossible to predict. Network integrity is therefore a control-engineering problem and one of the more important aspects of modern performance engineering.

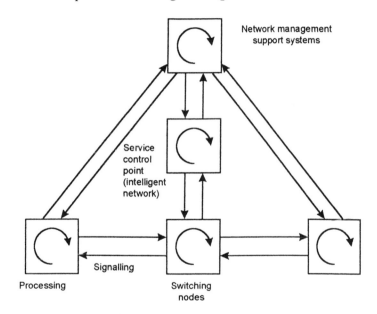

Figure 11.2   Network-control structure

There is an analogy to the personal computer (PC) on one's desk; almost everyone has experience a 'crash' on a PC due to a faulty application program or an unusual set of circumstances. But dealing with a PC is quite a different problem from that where a whole network has crashed. A major concern is the possibility that 'hackers' might introduce viruses into the network control systems, as they do with PCs [2].

A network is a collection of distributed entities performing different tasks and interacting with each other to provide the required service functionality. The behaviour of such a system can be represented as an extended finite-state machine [5]. The behaviour path is a set of states and the transitions between these, as illustrated in Figure 11.3. A path of correct behaviour consists of a sequence of acceptable states. Under adverse conditions, the system might leave this path and move to an erroneous state. Some errors will not produce a major deviation from the path of correct behaviour and the system remains in the 'shaded' area of acceptable behaviour. Such minor deviations can be classed as type 1 when they return to the correct path, type 2 when they are detected and cause the process to stop and type 3 where they terminate in a final state outside the normal path but still within the safe region. Conversely, major errors or combinations of errors can cause 'control mutations' which take the sequence outside of the region of safe behaviour and result in failures.

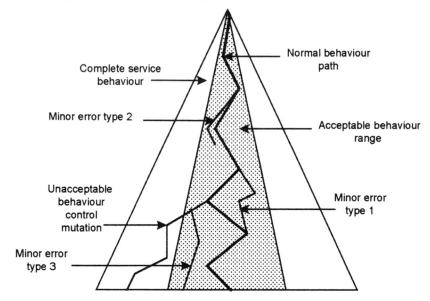

Figure 11.3   State transitions

11.4 Network interconnection

11.4.1 General

The risk of breach of network integrity is increased when networks are interconnected. This is becoming a major regulatory issue now that regulatory policy is allowing many competitive operators and service providers to interconnect to the networks of established telecommunications operators, e.g. the European Commission's ONP policy. Hence, it is necessary to consider network integrity in the context of network interconnection. The process is highly interactive as shown in Figure 11.4 [6] and starts with the market demand for value-for-money services needing specialised network platforms to carry and process the signalling and communications information. It is the transport and processing of this information which gives rise to the threat to integrity. The assumption is that the problems of integrity

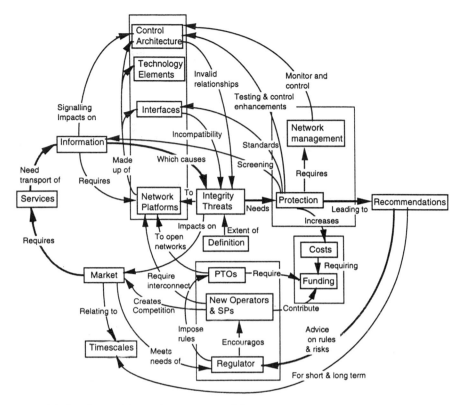

Figure 11.4   Network-integrity process model

are fundamentally concerned with network control, i.e. the transmission and processing of control information.

The ONP aims to open the networks of dominant telecommunications operators (TOs) to new competitors, i.e. service providers (SPs) or TOs, defined as other licensed operators (OLOs), to gain access to the TO's customer base or use the resources and services of TOs to offer their competitive services. Thus, an environment of extensive interconnected networks and SP equipment could develop. The networks of dominant TOs are interconnected to those of competing or specialised network operators (OLOs) to obtain total connectivity between all customers. Initially, OLO's gain access to the PTO's customer base to offer cheap trunk calls, sometimes known as trunk bypass; but OLOs are now competing by direct access to customers, e.g. cable-TV operators. Mobile operators connect to obtain connectivity to customers on fixed networks. Increasingly sophisticated private networks are also interconnected to the public networks.

In the future environment of highly-sophisticated intelligent network services, an increasing number of service providers will also require access to fixed and mobile networks to offer specialised services. A complex mesh of networks and management systems will therefore arise, as illustrated in Figure 11.5.

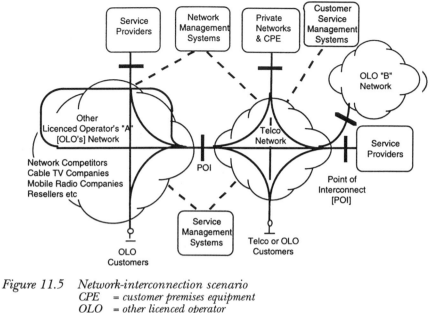

Figure 11.5   Network-interconnection scenario
    CPE   = customer premises equipment
    OLO = other licenced operator
    POI   = point of interconnect

The generic constituent parts of a network and its management are shown in Figure 11.6; traditionally they have formed the networks of the monopoly TOs. Combinations of the constituent parts are used to provide the network services. However, with the introduction of competition, new operators entered the market where the cost of entry was lowest and the potential for undercutting the established TO's price was highest. Initially this was the trunk-call market and the new operator provided a core transport network interconnected to the TO's network to gain access to the customer base of the TO. Cable-TV operators were able to provide access at a marginal cost and now compete in the access network connecting to the core transport network of other networks to route calls.

It can be expected that, in the future, SPs will offer value-added services from their own service platforms connected to the transport networks of other operators to gain access to customers and route calls. Information services will be supplied by independent information providers, particularly in the multimedia era. Hence, in an open network environment, customers will be able to gain services from a variety interconnected constituent parts of TO's and SP's networks and service platforms, as illustrated in Figure 11.6. The points of interconnect are determined by a variety of technical and economic factors.

The most common form of interconnection is related to the 'plain old telephone service' (POTS) where interconnection of public switched telephone networks (PSTNs) and mobile-cellular-radio networks to PSTNs

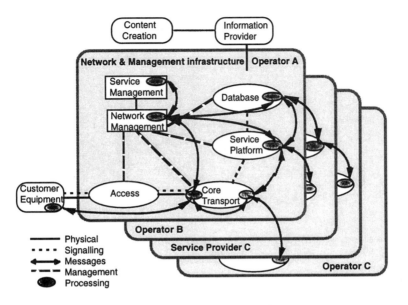

Figure 11.6   Interaction of constituent parts of networks

are already successfully operational using ITU-T 7 signalling. In the near future, the range of interconnected services will probably be extended by custom local-area-signalling services (CLASS)† services requiring the transmission of calling-line-identification (CLI) signals between networks. Intelligent services will follow with, for example, regulatory requirements for number portability between networks and the so-called CS 1 services, such as virtual private internetworking.

Network integrity can be compromised by problems migrating between interconnected networks and between equipments connected to them. The modern telecommunications network has a multiplicity of 'intelligent' switching and signalling elements interacting in real time. It is virtually impossible to test all possible interaction combinations; hence the possibility of failure is high. The position is exacerbated by the interconnection of networks containing equipment from multiple suppliers.

Currently, the most important threat comes with the interconnection of ITU-T 7 signalling networks where, for example, protocol or signal-transfer-point incompatibilities can cause major and widespread network disruption (sometimes called 'brownouts'). Likewise, the interconnection of synchronous digital hierarchy (SDH) networks will require particular attention to be paid to control-overhead aspects. It can be postulated that future co-operative intelligent services which require network-control interworking (network–network or network–SP) will be even more sensitive.

† CLASS services are provided by a number of network operators and include call trace, repeat dialling, CLI display, selective call ringing, selective call forwarding, automatic call back and selective call blocking.

Table 11.1 Integrity problems

Category	Problem	Effects	Cause	Risk	Cure	Comments
1	Signalling feedback loop	Congestion of signalling channel causing failure of interconnect link	(a) Protocol incompatibility causes repeat message requests (b) Overload controls continuously redirect messages between networks (c) Traffic congestion	High	Overload & loop-control procedures Traffic management	
2	SCP-pair failure	Disabling of links to other SCs causing failure of signalling network in on or both interconnected networks	SPT generic-software upgrade with faulty software	Med	Adequate testing	Cause of USA brownouts
3	Incompatible feature interaction	No interoperability of service(s) between networks	Application and/or protocol incompatibility	High	Mediation at interconnect Internetwork testing	
4	Gateway-switch recovery	Isolation of interconnect route, could oscillate as each end recovers	Mismatch of recovery processes at each end of interconnect link	High	Standard recovery process	
5	Timer incompatibility	Calls in set-up phase fail	Incompatibility of implemented timers, e.g. awaiting further digits	Med	Internetwork testing	
6	Incorrect message treatment	Node or call failures	Invalid treatment of messages at terminating gateway, e.g. due to erroneous data build	Med	Internetwork testing	

As increasingly sophisticated services are operated between networks, so signalling activity and interface complexity increases. It can therefore be concluded that there is an increasing risk of network integrity being compromised by problems arising from interconnected networks and equipment connected to the network. Examples of some types of problem which can compromise network integrity are shown in Table 11.1.

There is a broad correlation between the complexity of an interconnection, in terms of such aspects as signalling activity and the number and range of network elements that are accessed to provide a particular service, and the risk to integrity. For example, POTS would be low complexity but IN-based services, such as virtual private networks (VPNs), require a significant volume of transaction signalling to various databases. Clearly, conditions of interconnect will be influenced by risk and hence complexity. Thus the more complex services require more demanding interconnection conditions in terms of, for example, testing, screening etc. Therefore, to assist assessment of requests for interconnection it is desirable that a broad measure of complexity be developed for each class of interconnected and internetworked service.

Such a measure could, for example, be the product of signalling activity (in terms of messages per call) and elements accessed, measured against thresholds [4] to indicate high, medium and low complexity as illustrated in Table 11.2. However, work needs to be carried out to develop an appropriate index.

Table 11.2 Complexity measurement

Service	Signalling activity (messages/call=m)	Resources accessed (number=n)	Complexity m x n
Service A	a	l	$a \times l > y$ = high
Service B	b	m	$b \times m < z$ = low
Service C	c	n	$z < c \times n < y$ = medium

11.4.2 Interconnection of intelligent networks

The characteristics of future 'intelligent' networks (IN) [7] will be:

(a) an extensive set of network features providing a rich portfolio of customer services;
(b) the rapid development and deployment of new services;
(c) the ability to customise services for individual customers (perhaps under customer control) for sectors and segments of the telecommunications market; and
(d) extensive, regulator-driven, interconnection of networks of different operators and service providers with elements supplied by different manufacturers and requiring interoperability of services across the networks.

Service software will be placed in a variety of elements such as service control points (SCPs), intelligent peripherals (IPs), adjuncts and even switches themselves. Furthermore, services may be created by assembling a set of features into a 'feature package', using some features already developed and used in other feature packages. Potential interconnection possibilities for intelligent networks [6] are illustrated in Figure 11.7.

Feature interaction [8] occurs when the behaviour of one feature is altered in an unexpected and adverse way by another feature, hence preventing customers from obtaining the full benefits of specific services for which they have subscribed. Undesirable feature interactions may arise due to such things as incompatible feature functionality, protocol and logical ambiguities or insufficiency of available data. They may also involve logic on multiple network and CPE elements in interconnected networks, since implicit conventions may not be observed because software is developed, deployed and maintained by more than one organisation. Where undesirable feature interaction occurs on a service which requires network interoperability and prevents the service being offered between the networks or within a network, then network integrity has been breached. In the worst case, feature

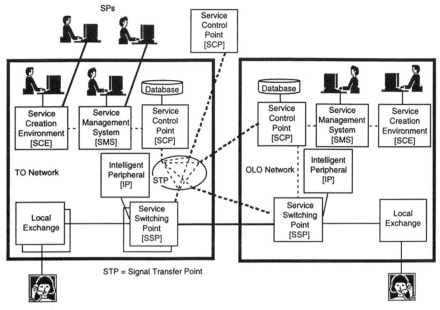

Figure 11.7   Interconnect possibilities for intelligent networks

interaction may cause catastrophic failure of basic services; for example, a feedback loop (i.e. repeated execution of one or more features) causes severe signalling overload. As the number of features increases, and there are already hundreds of them, predicting and detecting interactions becomes very complex and the risk of integrity violations increases. Ensuring protocol conformance at each interface is difficult because each developer may specify and interpret aspects of protocols differently.

The situation may be made more complex under an ONP regime. This permits open interfaces to third-party service providers at the application-programming interface (to execute their own service logic on the service-creation platform of the network operator) or at the application protocol interface (to permit interconnection to the ITU-T No 7 signalling network of the network provider). The potential problems are compounded under an ONP regime which permits open interfaces to third-party service providers to offer advanced services, perhaps using their own service-control point connected to the PTO's service-switching point. Such a regime is to be encouraged to stimulate innovative services and, provided that sufficient precautions are taken, restrictions on the grounds of essential requirements to protect network integrity should not be permitted. Indeed, extensive testing is being carried out in the USA to understand such interconnection and one Bell Operating Company has already publicly stated its intention to allow SPs to interconnect their own SCPs to its IN platform.

11.5 Measurement of loss of network integrity

11.5.1 The requirements

A broad definition of integrity is not precise, and there is a need to define network integrity in a manner which will allow network failure due to loss of integrity to be identified and preferably measured. The requirement to provide a quantitative definition arises because of:

(a)   the need to identify when a breach of network integrity has occurred;
(b)   the need to understand the magnitude of the loss of integrity in terms of the impact on network customers; and
(c)   the need to construct meaningful service-level agreements or contracts between network operators for network interconnection, or impose regulatory conditions which can be arbitrated in quantitative terms.

The Alliance for Telecommunications Industry Solutions (ATIS) T1A1 working group on network survivability performance has introduced a general framework for quantifying service outage from a user's perspective [11]. The parameters of this framework are:

(i)   the *unservability (U)* of some or all of the services affected by the failure, in terms of units of usage, e.g. calls;
(ii)   the *duration (D)* for which the outage exists;
(iii) the *extent (E)* in terms of the geographical area, population affected, traffic volumes and customer traffic patterns, in which the unservability exceeds a given threshold.

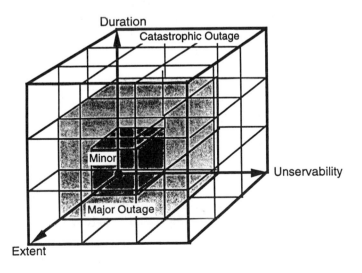

Figure 11.8   (U,D,E) qualifying regions

The service outage can be categorised by sets of values for which the (U,D,E) triple qualify for particular values of outage. This is illustrated in Figure 11.8 where the triples are categorised in three regions of minor, major and catastrophic.

The main problem in deriving an index is to quantify the impact on customers connected to the network(s), i.e. a local-exchange outage of x hours duration does not affect all customers connected to the exchange, given that only a proportion of them will attempt to make a call during that period. Questions which need to be addressed include:

(a) Loss of connectivity to the rest of the network merely affects those who attempt to make calls during the outage period, but what is the difference between partial and total isolation?

(b) What is the difference between a total outage and a severe degradation of service where, say, loss of network synchronisation causes slip with severe error performance which has a different impact on speech and on data?

(c) Is the impact on business customers greater than that for residential customers?

(d) What is the impact of loss of interconnection between networks? Is it different for fixed – fixed, fixed – mobile, PSTN – data etc.?

(e) How do you determine the impact on customers of different services? Does the tariff reflect the value of the service to the customers?

(f) How do you quantify the difference between a large highly penetrated network and a small low penetrated network?

Three possible measures of network integrity have been identified [4], but none adequately answers the above questions. They are the Cochrane 'Richter' scale, the McDonald 'user lost erlang' (ULE) and the ATIS 'outage index'.

11.5.2 Cochrane's 'Richter' scale

Cochrane has postulated that network disasters are analogous to earthquakes and can be categorised in terms of the Richter scale [9]. Total network capacity outage in customer-affected terms is thus:

$$D = \log_{10} (NT)$$

where N = number of customer circuits affected
T = total downtime

The qualitative ranges of media, regulatory and governmental reaction relative to this disaster scale are illustrated in Figure 11.9.

Advantages: The measure is easy to comprehend. It is impossible to derive an absolute value without having to specify outage in huge numerical values, and the logarithmic scale (like the Richter scale for earthquakes) reduces the numbers to manageable sizes.

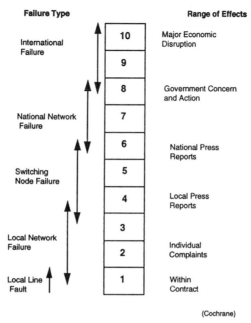

Figure 11.9   Network outage disaster scale

Disadvantages: For failures in the core of a network it would be necessary to convert, say, lost calls into affected customers, as is done for the ATIS index. It would need modification to deal with outages of important individual services with low traffic volumes.

11.5.3 User lost erlang

The parameter called user lost erlang (ULE) is defined [10] as:

$$\text{ULE} = \log_{10} (E \times H) \text{ for } E \times H \text{ greater than } 1$$

where E = estimated average user traffic lost during the time of
        the outage in erlangs taken from historical records

        H = outage in hours

Thus 1 ULE = 10 user lost erlangs, 2 ULE = 100 etc. Hence the unit is logarithmic (like the Cochrane Richter scale), as illustrated in Figure 11.10. It represents the societal impact in terms of calls affected and the outage duration, and it is easy to measure. It can also be translated into a measure of the estimated number of customers affected by dividing the erlangs by average calling rate in erlangs per customer line. The definition does not provide a measure of the geographical spread of the outage which, for a small network, could be extensive even though the value of ULE is modest.

Although loss of network integrity is commonly assumed to be the result of unavailability, it can also be the result of degradation of other performance parameters which affect large numbers of customers, e.g. poor error rate of a

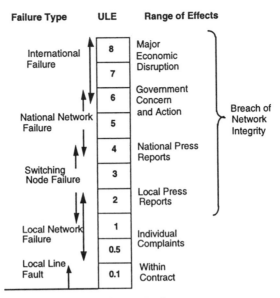

Figure 11.10   Definition and impact of integrity loss

service or services. In such circumstances, the ULE parameter could still be used as a measure of lost integrity by redefining E as 'the estimated average user traffic, in erlangs, lost or of unacceptable QoS during the time of the service disruption' and H as 'the service disruption in hours'.

The measure will indicate the greater impact on business customers over residential customers because of their increased calling rate.

Advantages: It is easy to calculate and comprehend. It handles customer and traffic outages. The use of erlangs reflects the actual loss of usage rather than the absolute number of customers

Disadvantage: It would need modification to deal with outages of important individual services with low traffic volumes.

11.5.4 ATIS index

The FCC in its February 1992 Report and Order challenged the telecommunications industry to develop a scientific method for quantifying outages. The Network Reliability Steering Committee passed the remit to the T1 Standards committee (both are committees of ATIS). Its subcommittee T1A1 proposed that the index should be based on a combination of service(s) affected, duration and magnitude (i.e. number of customers affected) [11]. 'In particular it should:

(a)   reflect the relative importance of outages for different services;
(b)   be able to be aggregated to allow comparisons over time; and
(c)   reflect small and large outages similarly to their perception by the public.'

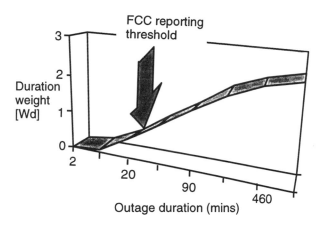

Figure 11.11 'S' curve for duration

An important objective of the exercise is to permit the summation of individual outage index values, over time periods related to the FCC reporting requirements (> 30 min, 30 000 customers), for comparison purposes. Aggregation should reflect the importance of outages to customers such that the aggregate index for multiple small outages is less than the index for one large outage whose duration (or magnitude) equals the sum of the small outages. Likewise, the aggregate index for multiple large outages should be greater than the index for one very large outage over the same measurement timescale (one year). This index behaviour is captured by an 'S' curve for duration and magnitude weights, as shown in Figures 11.11 and 11.12. This allows aggregates over time periods to be plotted and analysed using statistical trend analysis.

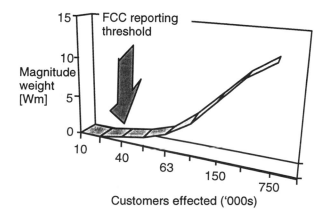

Figure 11.12 'S' curve for magnitude

The index of an outage is the sum of the service-outage index values for each service affected. The service-outage index of each service affected is the product of the:

(i)   service weight (W_s)
(ii) duration weight (W_d)
(iii) magnitude weight (W_m)

Thus, the outage index $I(O)$ for an outage O has the following form [4]:

$$I(O) = \sum_{j=1}^{N}(W_s(j),\ W_d(j),\ W_m(j)$$

where $j = 1,\ldots\ldots,N$ are the services.

Advantages: The method is simple to understand and operate. It is based on the FCC reporting requirements and is weighted to be used to aggregate outages over a time period. The method is promising and could be adapted and refined for use in a European environment. An integrity definition may be more acceptable to European operators and service providers if based on an acceptable practice in the USA.

Disadvantages: The assumptions are coarse and based on broad averages, e.g. a caller makes two reattempts when blocking occurs. It is currently used for a limited range of 'POTS' services. It does not differentiate between residential and business customers which have different perspectives of the importance of telecommunications.

11.6   Integrity risk

The network-integrity definition quantifies the extent of the effects of breach of network integrity when an incident occurs. The categorisation of a network or service outage was the consideration of Section 11.5. Even in this context it has been pointed out that further work is necessary to characterise transportable agreed definitions. Specifically, the details of the weighting functions need careful examination. In addition, there is a need for an agreed framework for expansion of the definitions to encompass new services. The ATIS outage index offers this framework.

However, there is a further development which is necessary but cannot easily be contained in the present structure. Once it is possible to quantify network outages, it becomes desirable to quantify threat of outage as a consequence of some action, e.g. due to failure of a specific type of interconnection or internetworked service. In fact, this should be regarded as the real definition of integrity. This is extremely difficult and is not proposed as a necessary requirement to the introduction of interconnection. However, it has been identified as highly desirable and every effort should be made to support its development.

The definition and development of such a measure of robustness is intimately related to modelling and development of interconnect testing. Much of what has been discussed has a direct bearing on the requirements here. For this reason the following will be quite brief, but will indicate where a connection to other sections is relevant.

There are major difficulties in trying to obtain a quantitative measure of a threat to a network. This is due in part to the nebulous nature of the term 'threat', but more to lack of quantitative understanding of network behaviour. Therefore, proposing that a given interconnection process poses a threat to the integrity is no more meaningful than saying that it does not. What would be desirable is to say that a given interconnection increases the likelihood of a network failure by a probability x. Hence, it would be possible to then say that if x exceeds an agreed threshold value, it constitutes an unacceptable threat. To achieve such a numerical measure may not, in absolute terms, be possible; but an indication of the threat within error bounds should be. The development of such measures will only be possible from experimentation, run in conjunction with modelling. This is because the essential process is predictive; therefore there has to be some theoretical basis to achieve this.

Digital exchanges and signal-transfer points are subject to regular changes in data and upgrades to software design. Hence, interconnected networks are in a state of constant change which may invalidate previous interconnection testing. Furthermore, although standard interfaces may be assumed, in multivendor interconnected networks each manufacturer may interpret standards differently. Experience in the USA has shown that corruption of data and changes to software are a significant cause of intranetwork outage, particularly for ITU No 7 signalling networks; it can therefore be assumed that there are similar internetwork risks. Indeed, some operators insist on a reduced test programme being carried out when significant software updates are made at the point of interconnection.

In an environment of multiple-provider interconnected-networks, it may be assumed that interconnection interfaces should be defined by standardised protocols, such as ITU No 7. This implies that the individual software elements within the networks can be designed by different manufactures and administered by different network operators. However, conformance to standards can still lead to incompatibility because each manufacturer may interpret protocol aspects differently. Given the continuous upgrade of software, compatibility between generic levels of software across interconnected networks becomes difficult to maintain.

Ideally, the assessment of a risk should be met by an assessment based on a model which would assess the level of risk to integrity and the threshold criteria would determine its acceptability. Such models do not exist, but an approach is illustrated in Figure 11.13 [12]. The initial phase is to describe the service in terms of the message interchange between entities and their consequent actions. This will enable the correct behaviour sequence (shown in Figure 11.3) to be identified. It is then necessary to postulate possible error events, i.e. perturbations such as invalid messages that may trigger a violation of integrity, and to study how the introduction of these errors affect the behaviour of the service. Two inputs are required to identify errors, namely:

(a) Performance models: These represent errors of a *static* nature such as bad definition of services, poor logic, incorrect programming etc., and *dynamic* errors due to performance limitations from delays, signalling overloads etc.

(b) Customer models: The behaviour and operation of a service is strongly effected by the way the service is used by the customers. These customer models can be decomposed into two main areas, called *user models* and *population models.* The user models represent how an individual user would interact with the service, e.g. what actions they are likely to take, possible errors they would make and how they would react in different situations. The population models represent the characteristics of the whole collection of users of the service. They include, for example, the number of customers, their geographical distribution, their degree of mobility, and their usage patterns (e.g. call-holding times, average and peak calling rates etc.).

The message-sequence charts [13] which describe a normal path of behaviour need to be modified to introduce errors. Analysis of this information will permit the identification of error states which can be

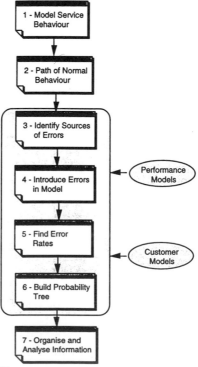

Figure 11.13   Predictive modelling

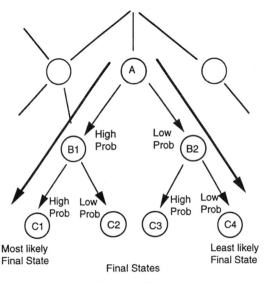

Figure 11.14   State transitions and probabilities

incorporated in the model in order to build paths of erroneous behaviour. After a list of the possible error events has been identified, they must be matched to rates, i.e. errors occurring per time unit. Performance characteristics and population models need to be taken into account to obtain the probabilities of each particular type of error. Initially, error rates must be assigned based on external information, e.g. previous experiences, and there will be an associated degree of uncertainty. This uncertainty will always exist, but it will be reduced as the methodology progresses and provides more information about errors and their probabilities.

The worst problems are likely to arise when a chain of errors occurs; hence the need to identify possible sequences of error states which will progressively take the system away from the normal behaviour. The correlation between different error states, i.e. the probability that one error state leads to another, forms a large and complex decision tree. A simple example is demonstrated by the tree structure illustrated in Figure 11.14 [5]. Each node represents a state and the interconnecting branches represent transitions between states and their probabilities. This will show the overall probability of moving from the initial state to each of the possible final states.

In complex systems, where the number of possible paths and final states is extremely large, it would not be feasible to find the probabilities to reach all of the states. Therefore, the work must concentrate in two cases, namely:

(a)   the most-likely states (such as state C1), which describe the normal operation of the system; and

(b)   the most dangerous states, i.e. those which fall outside of the safe region, even if their occurrence is very unlikely, because they need to be identified quickly in order to prevent catastrophic failures.

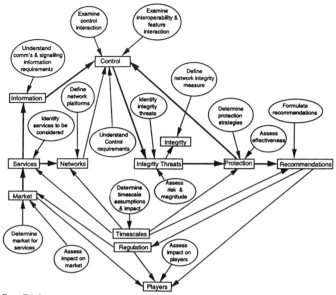

Figure 11.15   Risk-assessment process

Regardless of the improbability of entering a dangerous control sequence, identification of such sequences is extremely important. This is because a system operating for a significant length of time will explore a considerable fraction of all the possible sequences. Hence, it will eventually enter this dangerous region with catastrophic consequences. In summary, unlikely events are inevitable, provided that enough time elapses and, if such unlikely events lead to a threat to integrity, the threat will manifest itself. The key point is to identify them rapidly and take the appropriate actions before the system falls over.

The last stage of the methodology is to extract useful information from the data obtained in the previous stages, so that in can be easily accessed and utilised as a pre-emptive framework. There is the need for adequate documentation of all the acquired information, possibly in the form of a new kind of expert system where it can be extracted in an intelligent way.

A more pragmatic and broad-brush methodology is illustrated in Figure 11.15. However, it is suggested that a decision-tree approach be taken which gives a view of the degree of complexity and hence risk. This approach is illustrated in Figure 11.16.

11.7 Development of new services

The process from inception to launch of a new service [14] is illustrated in Figure 11.17. The process starts with a feasibility study on which the requirements specification of the overall system are based. The architectural

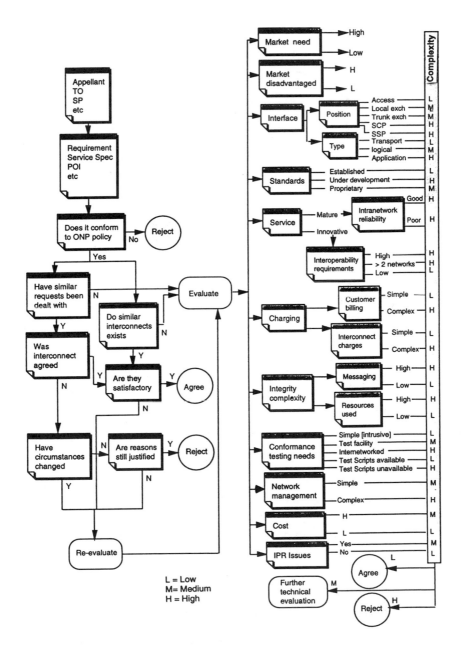

Figure 11.16 Assessment decision tree

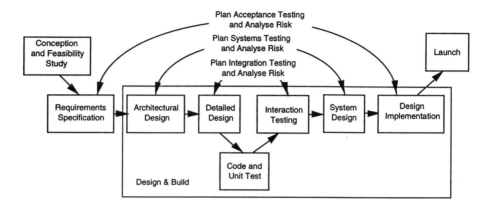

Figure 11.17   Development of new services

design sets the parameters within which the detailed design is carried out and the different components of the architecture and their interoperability are considered. This is followed by the most detailed level of unit design and coding. Each of these processes requires a different level of testing. At the most detailed level, testing must ensure that the individual coding works. Then, interaction testing is necessary to ensure that when two or more components are put together they perform as expected. When the overall system design is completed, acceptance testing is carried out to ensure that not only does the system work but it also conforms to the requirements. The launch phase requires not only testing in an operational environment, but also trialing of the in-service support processes.

At the customer-requirements-definition stage, consideration should be given to how the service will be used by customers, in particular the potential for undesirable feature interactions with other services. Also, information should be provided to enable population and user models to be constructed and the likely QoS requirements to be determined.

Early in the design stage, modelling should be carried out to assess the risk of feature interaction and integrity violation. The objective is to remove problems at the early stages of the design cycle, to avoid the time-consuming and costly method of identifying problems when the development is completed. At present, services are more often designed in isolation, without taking any account of existing services, and therefore may give rise to interworking problems.

Testing should be carried out at appropriate stages in the development, as shown in Figure 11.17. Acceptance testing should embrace not only testing of the completed design on laboratory models but also intranetwork field testing of the installed equipment and software. Where the service is to be internetworked via the networks of other operators, comprehensive non-intrusive testing should be carried out between captive models followed by intrusive testing of the internetworked services, in the operational environment, before they are used by customers.

Owing to the high complexity of the services, it is not possible to test everything. Testing must focus on those areas where problems are most likely to arise, whereas aspects that can be assumed to be nondangerous will not be included in the testing. The level of testing carried out must also be subjected to risk assessment. Since it is clear that these limitations will exist, it is of paramount importance to understand their consequences. For this reason, a risk-assessment methodology which helps to assess the probability, impact and consequences of actions must be developed and integrated into the development of systems. This can be viewed as an addition to the development cycle shown in Figure 11.17. This figure demonstrates that risk analysis should be applied at the different levels of systems development, in order to determine the amount of testing that needs to be performed for different levels of risk. Increasing the amount of testing decreases the risks, but it increases the costs and delays; so a balanced approach must be taken. The risks should be analysed, categorised and documented.

Part of the risk-assessment activity should consist of keeping records of any constraints decisions taken during the development of the system, and an assessment of the effects that these limitations would have in adverse conditions. Such an activity should:

(a) provide a better understanding of the systems operations;
(b) make the information accessible to different people, instead of being only in the heads of those involved in the design process;
(c) help to identify weak points in the systems and where problems are likely to arise;
(d) help in the diagnosis of failure and lead to a rapid identification of the actions required to fix the problems; and
(e) help to evaluate consequences of failure, e.g. in terms of lost calls, damage to equipment, cost etc.

The areas of development work related to integrity can been classified in four main groups, and each of these into some subgroups. The four main functional groups are:

(i) *Service provision:* This includes all activities involved in the provision of a new service, from the requirements capture and service specification to the design, implementation and testing of the service. Testing has been included as a separate activity at the end of the development cycle, for convenience, but it is important that it spans all the steps of the development cycle of any system.

(ii) *Interconnect:* This covers activities dealing with the design of policies and testing for interconnect to other operators.

(iii) *Intelligent networks:* This is a relatively new area and it is still at a research stage. At present, most services provided by Telcos are POTS and some ISDN, but the tendency is towards IN-based services. There is increasing research activity in this area, both in the provision of IN-type services and in the construction of the appropriate platforms to provide them.

(iv) *Signalling network:* The provision of new services often has implications for the signalling network, if only to increase its loading.

Note that these activities must not be treated in isolation since there are strong connections between most of them. For example, an IN type of service obviously falls into the categories Intelligent networks and Service provision, but also, if the service is going to be provided globally, this means that Interconnection issues need to be considered. Finally, owing to the strong relationship between IN and noncircuit-related signalling, details about the Signalling network which is going to convey all the signalling information for the service will have to be taken into account. Hence, it is important to appreciate that the aim of this functional decomposition is to simplify the overall view of activities related to the integrity problem. Other examples of areas which fall into more than one category are interconnect-signalling policy, service interaction and regulation.

Other activities and factors which should be considered in order to tackle the network-integrity problem are:

(a) *Population and user models:* The behaviour and operation of a service is strongly affected by the way the service is used by the customers.

(b) *QoS:* Quality of service represents how the service is perceived by the customers. In a market environment such as telecommunications, where customer satisfaction is a primary objective, special care must be taken to understand those parameters that measure customer perception and to maintain the appropriate values for them.

(c) *Network performance:* The behaviour of a service can be influenced heavily by network-performance characteristics. Therefore, the study of possible erroneous behaviour must include its relationship with performance.

(d) *Signalling:* This factor refers to the general information flows within the system, as opposed to specific signalling protocols or architectures, represented under the work area 'signalling network'. Signalling is the area of most growth in complexity, and hence highest risk to integrity, in the new telecommunication scenarios. For this reason, it is a factor which must always be carefully analysed.

(e) Monitoring: This is needed to check progress throughout the development cycle and in service, where it is important to detect potential integrity violations before they affect customers.

(f) Automated records: As systems evolve and become more complex, the amount of information which is produced in their development and is required to be understood increases. Thus, the traditional manual approach to the maintenance of network integrity, which relies on the knowledge of individual experts, is no longer valid. Hence, it becomes necessary to store this information in a structured and automated way which makes it accessible, meaningful and useful to different people.

(g) Risk analysis: Part of the problem of maintaining network integrity is being able to identify areas of risk, and to predict the consequences of limitations and constraints of the developed systems. Hence, we believe that risk-assessment methodologies should play a main role in tackling the integrity problem.

(h) Modelling: Modelling and simulation techniques are needed to improve the understanding of the new systems in the early stages of development. This is particularly important in the telecommunications environment, which is characterised by fast changes and increasing complexity.

The relationships between the above factors and the four functional areas can be presented in the form shown in Table 11.3, which represents the relationship between the different work areas of the network operator and the factors that have been identified as relevant to the integrity problem. The content of each square shows whether the factors on the left relate to the functional areas at the top.

11.8 Integrity protection

There are a number of ways to protect the integrity of networks when they are interconnected. Some of the more common ways are:

(a) Gateway screening: Software, usually embedded in gateway exchanges at the points of interconnection, contain a 'mask' of legitimate messages and therefore reject all others as invalid. This is used by many operators for POTS interconnect.

(b) Firewalls': The interconnection of more complex services, for example, transaction signalling, may require more comprehensive protection, such as a 'firewall'. Typically, this may be in the form of a signalling relay point which, in addition to containing a mask of valid messages, can block messages to invalid signalling destinations (point codes). It can also change signalling routing to avoid network 'hot-spots' and act as a fuse when unexpected signalling surges arise to prevent them from propagating into the network.

Table 11.3   *Typical mapping of areas of work to activities and subjects related to integrity*

	Service provision					Interconnection		IN		C7 network	
	Requirements	Specification	Design	Implementation	Testing	Policy	Testing	Services	Platforms	Design	Maintenance
Population models	!	!	!	!	!	!	–	!	!	?	?
User models	A	A	A	A	A	–	A	A	?	–	–
QoS	?	?	?	?	A	A	A	?	?	?	?
Performance	?	?	?	?	?	A	A	?	?	?	?
Signalling	–	–	A	A	A	A	A	A	A	A	A
Monitoring	A	–	–	–	A	–	A	?	?	A	A
Automated records	?	?	?	?	!	!	!	!	?	?	?
Risk analysis	!	!	!	!	!	!	!	!	!	!	!
Modelling	–	A	A	A	A	!	!	?	?	–	?

A activity carried out
! activity not carried out
? lack of information
– not applicable

(c) Mediation devices: Whereas screening and 'firewall' devices are used for blocking unwanted messages, mediation devices modify the content or structure of messages to achieve compatibility between source and destination. Such devices will probably be required for interworking of complex IN-based services where interconnection is, say, at the SCP level.

11.9 Interconnection testing

11.9.1 Testing requirements

In order to achieve seamless service operation with high-integrity protection, a sequence of testing programs must be carried out which encompass stand-alone technical analysis and auditing of individual systems and multisupplier interoperability, as well as internetwork interoperability. The depth of testing will obviously depend on the complexity of the service and interconnection. Responsibility for these tests should be shared between the equipment vendors, the service providers and network operators, as shown in Figure 11.18.

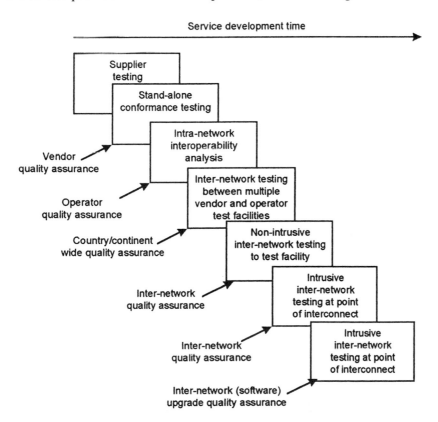

Figure 11.18   Testing sequence

The design for the equipment is usually in accordance with industry standards and the individual specifications of the client operator.

(a) The first step in the technical analysis is a thorough design review of the specifications. This concentrates on potential weaknesses, such as overload-control algorithms, which have been identified previously. It usually becomes prohibitively expensive to correct a fault not detected at the design stage.

(b) The second stage involves stand-alone conformance testing on each system to verify the correct implementation of the specifications. This includes a check on the generic requirements and SS7 standards, capacity and performance, security, hardware and software quality, and, in particular, the feature and service capabilities. Any corrective actions can thus be performed prior to product release.

(c) The third stage of testing considers interoperability necessary within the operator's network, often between multivendor systems, for transparent service operation. Such tests are usually carried out within the operator's test facility before intrusive tests are carried out in the network.

(d) Ideally, if network interconnection is likely for internetworked services, the next stage of testing should be co-operative, between the interconnected test facilities of a number of vendors and operators. For such multivendor/operator situations, diversity in the interpretation and implementation of the specifications can lead to operational difficulties. Such testing could take weeks rather than days.

(e) Stage five arises when an application has been made for a specific interconnect, say, from a new entrant to an established PTO. If the switch is unfamiliar to the PTO, it would be reasonable to insist on interconnection testing to the PTO's test facility. In addition to straightforward interoperability tests, stress testing should be carried out to observe the reaction to such things as overload and fault recovery. Such tests and analysis could take four or five weeks.

(f) At time of interconnection, intrusive tests should be carried out at each point of interconnection between the two 'live' networks. Such tests would typically comprise electrical and alarm tests together with test calls and SS7 (ITU-T SS 7) signalling tests on each type of route (e.g. emergency, assistance, egress, ingress etc.) with diagnostic testers to check the message sequences. Typically, this might take three or four days for an experienced operator. Such tests may be reduced if similar equipment has been used without problems by other operators interconnecting to the PTO.

(g) If a new software generic type is introduced or additional capacity or new routes are added to the interconnection a reduced set of intrusive tests should be carried out.

There appears to have been no recorded incidence of a breach of integrity caused by the interconnection of networks or connection of service provider

(SP) equipment to networks. However network interconnection to date has been for relatively simple voice (POTS) services using well defined national-standard SS7 interfaces. Additionally, such interconnections usually have some form of gateway screening by PTOs to isolate messages corresponding to invalid relationships. PTOs also require substantial interconnection testing before they permit connection to their networks.

Interconnection testing can be time consuming and expensive. Unlike customer-premise-equipment type approval, which is carried out by neutral bodies, interconnection testing is dictated by the dominant PTOs. It is conceivable that PTOs could impose over-elaborate testing which would discourage or delay other operators or SPs from connecting to their networks. In the USA, the telecommunications industry organisation ATIS has produced standard and agreed test scripts, although their use is not mandatory.

In the future, it is conceivable that sophisticated internetworked services may require the collaboration of resources in more than two networks. Hence, even though bilateral interconnection testing is satisfactory, the use of three or more networks to provide an internetworked service may give rise to integrity violation in one or more networks.

11.9.2 Interoperability test facility

In the USA, where the potential problems of network integrity are well appreciated, significant internetwork testing is taking place between interconnecting vendors' and operators' test facilities via the Bellcore

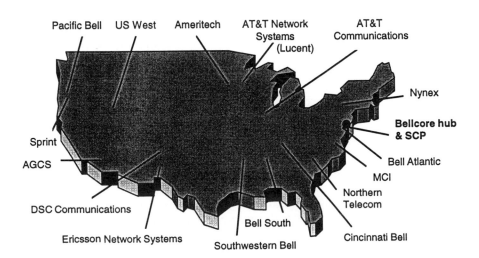

Figure 11.19 Interoperability testing

Network Services Test System (NTST) which monitors and analyses combinations of interconnected equipment types and networked configurations.

The Interoperability Analysis Program [4], initiated as a co-operative venture between the suppliers, RBOCs (Regional Bell Operating Companies) and Bellcore in the USA, has the aim of revealing those potential interoperability problems which are not identified during stand-alone testing. A controlled environment is created, since the suppliers' laboratories are directly connected via dedicated links to the Bellcore test bed, which contains all possible elements that could be encountered in a live environment (see Figure 11.19). This hubbing arrangement allows the replication of any specific network configuration. Bellcore's NSTS is used to monitor the simulated network, generate signalling-message traffic and collect test data. All participants are actively involved in the development of test plans, which focus on network integrity and service integrity. Bellcore is responsible for co-ordinating the simultaneous execution of tests by these participants.

Test scenarios include realistic network-failure conditions, such as link set failures, and the investigation of interaction among service switching points (SSPs) and service control points (SCPs), signal transfer point STP–STP compatibility and traffic-overload situations, which could develop when excessive queries are directed to individual SCPs (e.g. by media-stimulated demand for a network-portable 800 number). In addition, recovery procedures and network-management controls, such as automatic call gapping, are monitored through the NSTS. According to Bellcore, this testing procedure has been instrumental in protecting the integrity of the network.

Although it is difficult to provide figures for financial justification of the scheme, it is argued that one serious outage has such an enormous economic impact that it makes the cost of testing seem trivial. Example findings include incorrect coding of messages which causes their rejection, and incorrect distribution of signalling-link selection codes, causing unbalanced traffic. One technical analysis of a switch upgrade identified some 300 faults, feature interactions and other glitches during the test program, any one of which could have a detrimental effect on the normal operation of the network. However, it is impossible to test for all possible interactions which may occur in live operation, and some bugs will inevitably remain undetected before the systems are put into service.

11.10 Conclusions

The most severe degradation that customers can experience is loss of access to the network due to its failure; the greater the network outage the larger the impact on customers. The modern digital network is, in effect, a very large distributed-processing system having multiple varieties of exchange processor with a multiplicity of software operating systems and service applications. The complexity of such a system is high and it is rising with the growth of services and network interconnection between operators. Although interfaces are specified by standardisation bodies, it is not possible to produce a detailed unambiguous specification. Hence, realisation of standards depends on the interpretations of equipment suppliers. Therefore, the risk of incompatibilities is high and the probability of software-driven network outages is large. Furthermore, events in the past have shown that even minor errors in software design can cause major service-affecting network problems when unexpected combinations of circumstances or perturbations occur. There is also a long history of 'hacking' into computer networks and the malicious insertion of viruses which can cascade through computer systems rendering them inoperable. There is no reason to assume that telecommunications networks are immune from such attacks.

It must be recognised that modern networks are fragile and that the design of networks and systems to maximise the preservation of network integrity is of paramount importance. It will have an increasing influence on customer-perceived QoS, particularly as networks become more complex and internetworked services become more sophisticated.

11.11 References

1 US CONGRESS: 'Asleep at the switch? Federal Communication Commission efforts to assure reliability of the public telephone network'. Fourteenth report by the Commission on Government Operations, House Report 102 – 420, US Government Printing Office, 1991
2 WOHLSTETTER, J. C.: 'Gigabits, gateways, and gatekeepers: reliability, technology and policy', *in* LEHR, (Ed.): 'Quality and reliability of telecommunications infrastructure' (Lawrence Erlbaum Associates, 1995)
3 COMMON CARRIER BUREAU: 'Preliminary report on network outages', FCC 1991, (there was no final report)
4 WARD, K., *et al.*: 'Network integrity in an open network provision (ONP) environment'. Final report of study for the Commission of the European Union (University College London, 1994)
5 MONTÓN, V., WARD, K., WILBY, M., and MASSON, R.: 'Risk assessment methodology for network integrity', *BT Tech. J.*, 1997, **15**, (1), pp. 223–233
6 WARD, K.: 'The impact of network interconnection on network integrity', *J. Inst Br. Telecom. Eng.*, 1995, **13**, (4), pp.296–303
7 ITU–T, draft Recommendation Q.1221: 'Introduction to intelligent network capability set 2' (International Telecommunications Union, Nov 1995)
8 CAMERON E, J., and VELTHUIJSEN, H.: 'Feature interactions in telecommunications systems', *IEEE Comm. Mag.*, 1993, **31**, pp18–23
9 COCHRANE, P., and HEATLEY, D. J. T.: 'System and network reliability' (Chapman & Hall), chap. 11, pp. 6, 196
10 MACDONALD J. C.: 'Public network integrity – avoiding a crisis in trust', *IEEE J. Sel. Areas Commun.*, 1994, **12**, pp. 5–12
11 ATIS: 'Technical report on analysis of FCC-reportable service outage data'.(Alliance for Telecommunications Industry Solutions, 1994), document T1A1.2/94-001R3
12 MONTÓN, V.: 'An approach to tackling network integrity'. *Third Communications Networks Symposium*, Manchester Metropolitan University, UK, July 1996
13 ITU Recommendation Z.120: 'Message sequence chart (MSC)' (International Telecommunications Union 1993)
14 MONTÓN, V., WARD, K., and WILBY, M.: 'Maintaining integrity in the context of intelligent networks and services'. Fourth international conference on *Intelligence in Services and Networks*, Como, Italy, May 1997

Chapter 12

Economics of quality of service

12.1 Introduction

There are both costs and benefits associated with the provision of an acceptable QoS, but little work has been done to quantify the economic dimensions of QoS and its treatment is usually subjective and fragmented. It is not possible accurately to quantify the economic dimensions, which are often soft and speculative in nature. Nevertheless, it is necessary, at least, to carry out a qualitative cost–benefit analysis when introducing QoS initiatives to gain an understanding of the possible economic consequences. This requires an understanding of the economic drivers and their linkages, which is the subject of this chapter.

The costs and benefits can be progressively expanded as shown in Figure 12.1. The main categories of cost are those associated with the failure to achieve a satisfactory level of QoS, i.e. failure costs; appraisal costs which relate to activities concerned with the assessment of QoS; and those necessary to ensure that a desired level of QoS is obtained (i.e. prevention costs). The costs can be further decomposed into those directly effected by the network and the others. Benefits usually accrue in terms of increased revenue. In a competitive environment this is often the result of minimising loss of market share to competing operators and service providers. This is particularly important to the incumbent operator which finds it difficult to compete on price because of the burden of the 'universal service obligation' to serve all customers at a fair price. Such operators must therefore differentiate themselves from their competitors by the QoS they achieve.

12.2 Cost of quality

The traditional curves associated with failure and prevention costs are illustrated in Figure 12.2 [1]. The summation of the two gives the total quality costs, which would appear to indicate a minimum cost which provides optimum quality and differs from the zero-defects level. However, this ignores the benefits aspects of QoS and may not be the most profitable solution.

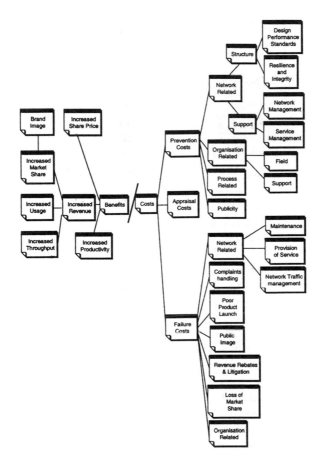

Figure 12.1   Quality-of-service costs and benefits

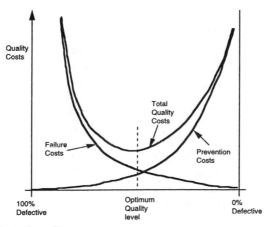

Figure 12.2   Cost of quality

12.3 Prevention costs

12.3.1 Definition

Prevention costs are those which are necessary to ensure that the products and services delivered to the customer, i.e. the offered quality, meets an acceptable level of quality. They include costs related to the structure of the network as well as the costs of all activities required to prevent production and delivery of substandard service offerings.

12.3.2 Network prevention costs

Prevention costs associated with the network are primarily related to its design and management. The three main parameters which affect the design and dimensioning of the network are cost, performance and throughput. Generally, the design of networks is optimised for minimum cost for a prescribed throughput and performance level, although any of the three variables could be optimised with the other two fixed. Chapter 10 showed how many of the performance parameters mapped onto the network-related QoS elements; also, that the end-to-end performance for each parameter, and hence the offered QoS, was broadly the summation of the individual performance of each link and node making up the connection, as illustrated in Figure 12.3. For many performance parameters, there is a trade-off between the cost and quality for each link and node. Hence, it is possible to arrange for the apportionment of the target end-to-end performance between nodes and links in such a way as to minimise the cost of achieving it. It is generally most costly to obtain good performance in the local access network (local loop) because of its low-grade technology, hostile environment and distributed nature. Therefore, the most stringent performance targets are allocated to the nodes and links in the core of the network where traffic is most concentrated and performance most easily controlled. For each node and link, costs and performance tend to be inversely proportional, as shown in Figure 12.3.

The use of equipment with less stringent mean time between failures (MTBFs) may reduce capital-cost requirements. However, in order to meet customer's required availability targets, more maintenance effort (and hence more operating costs) may be needed. Thus, there is a trade-off between capital and operating costs to meet a given availability target, and the optimum should be to minimise whole life costs, i.e. the combination of capital costs (represented by their equivalent annual charges) and operating costs.

In a similar fashion, there is a trade-off between capital (prevention) costs and operating (failure) costs for maintenance and provision of service in the local loop.

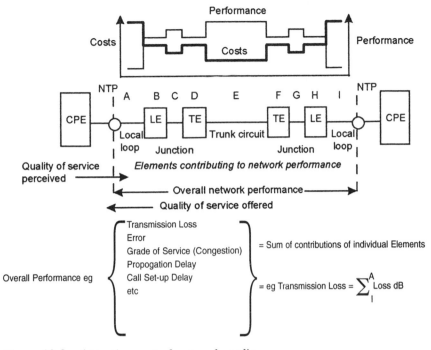

Figure 12.3   Apportionment of cost and quality

        CPE   customer-premise equipment
        NTP   network-terminating point
        LE   local exchange
        TE   trunk exchange

Traditionally, capacity is provided in increments to meet growth for a period of time (the design period). However, owing to the difficulty in forecasting growth at this distributed periphery of the network, provision of service often requires manual intervention at the flexibility points, i.e. primary crossconnect points (PCPs) and distribution points (DPs), to utilise available pairs. In addition to incurring failure costs in service provision, this intervention and disruption of the network creates faults, thus increasing the failure costs of repair.

In an extreme case, stringent QoS targets for time to provide service can lead to expedient provision, such as pair diversion, due to the nonavailability of a pair when a new customer requires service. This is not only expensive, but it diverts effort from normal provision of plant to meet the growth forecast, lengthens the provision-of-service time and worsens the fault rate. Such reactive planning to providing service reduces effort available to provide plant to meet forecast requirements, hence increasing the shortage and reducing the probability of pairs being available when and where required. Such a cumulative situation can rapidly run down network capacity and it is exacerbated by tight staffing levels and short provision-of-service targets which may influence manager's pay.

This problem may be overcome by providing sufficient capacity to provide for all unserved properties, thus making the network independent of forecasts. Such a 'stabilisation' of the local loop can incur high (prevention) capital costs; however, provision of service no longer requires physical intervention, thus reducing failure costs. Prior and Chaplin [2] demonstrated that stabilising the local loop in a particular local exchange area (St. Albans in the UK) had a significant impact on reducing the fault rate. They argued that the whole-life fault liability of local loop plant follows the traditional 'bath-tub' curve shown in Figure 12.4a and that when a cable joint or cross-connect point is opened and physically altered to provide service the whole of the particular

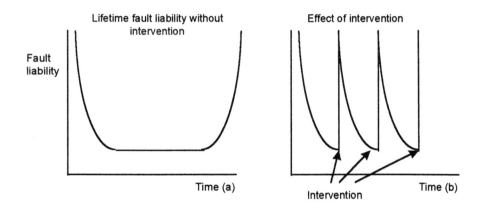

Figure 12.4   Effect of intervention on fault liability

element (not just the pair) could be moved back to the start of the 'bath tub' giving the 'saw-tooth' reliability curve shown in Figure 12.4b.

Additionally, if the unserved properties are preconnected to the local exchange and given a class of service (soft dial tone) which only allows calls to emergency services and the network operator's sales office and bars incoming calls, service can be immediately provided by the customer plugging in a telephone and calling the sales office. This dramatically improves the QoS for service provision.

12.3.3 Network resilience

To minimise disruption to service due to equipment failure or traffic overload, resilience is often built into networks. This is because an efficient network design reduces spare capacity and hence increases susceptibility to traffic overload. The ability to provide service under overload conditions is known as network resilience (or network reliability/security). The effect of network overload is increased with the use of common control exchanges and automatic alternative routeing (AAR). Then, congestion causes increased control activity and hence delay which, in turn, gives rise to repeat attempts; thus the cumulative effect can cause rapid degradation in service. This can cause blocking on outgoing routes from a particular exchange which would not normally be in congestion. Although AAR is valuable in absorbing small overloads, it can, in itself, spread congestion through the network under severe overload conditions unless carefully controlled.

The main causes of overload are the reduction in network availability (i.e. the probability of free paths through the network) due to equipment breakdown, the effects of which become more pronounced due to the increasing trend for higher-capacity transmission systems and larger, more complex exchanges; and traffic surges due to the increasingly sophisticated usage of the telephone by subscribers, e.g. phone-in radio and TV programmes etc.

Methods of improving network resilience are:

(a) improving reliability of components and systems by design and manufacture;

(b) adopting network structures less sensitive to surges and equipment breakdowns, e.g. grid and mesh networks;

(c) network redundancy by oversizing of capacity of nodes and links;

(d) provision of a 'service-protection network', i.e. spare transmission links which can be substituted for faulty traffic-carrying links;

(e) diversity in the physical routing of circuits making up specific traffic routes;

(f) looped-transmission-network configuration, where circuits are routed both ways around the loop to preserve capacity should the loop be cut by failure of a transmission system;

(g) controlled automatic alternative routing of traffic to avoid congested or failed nodes or links;

(h) load partitioning of switching nodes, i.e. major nodes are divided into two self-contained switching units with incoming and outgoing traffic routes divided between the two; and

(i) real-time network management to monitor network performance and take action to overcome congestion by rerouting of traffic, blocking traffic surges at their source exchanges and initiating repair of equipment faults.

All these measures involve expenditure and it is therefore necessary to select a combination which meets the needs of the network operator at an acceptable cost. This can be to provide maximum network-traffic capacity at a given cost and resilience standard, or to minimise cost for a given capacity and resilience standard, or maximise resilience for a given cost and capacity. It should be appreciated that the requirement for network resilience is a relatively new planning parameter and much research needs to be carried out to develop criteria against which resilience can be measured and procedures to optimise networks for resilience.

12.3.4 Network integrity

The increasing threats to network integrity due to the interconnection of networks and their increasing complexity and the provision of sophisticated services are described in Chapter 11. The economics of integrity can be grouped into failure and prevention. There are also commercial implications for customers and service providers of denial of interconnect.

The cost of failure of network integrity needs to be evaluated for the various failure modes. These can be:

(a) catastrophic failure, where large outages occur such as failure of a switching node or nodes, and can be very costly in terms of lost revenue. This cost can, in theory, be evaluated if the number of affected customers is known together with their average traffic (and average charge per call) together with the duration of the outage. However, this does not take account of repeat calls after restoration of service, or contractually binding rebates and litigation. Neither does it take account of loss of customer goodwill and confidence which may result in the loss of customers to other operators;

(b) failure of interconnect, where all calls passing between the networks are prevented from completing. Again, theoretical lost revenue can be broadly calculated and would affect both the operator in whose network the calls originated and that in whose network the call terminated;

(c) failure of a particular service (or services) in one or other network, or just those which require interoperability between networks. Again theoretical revenue loss can be calculated;

(d) degradation of performance which does not effect call completion, e.g. high error rate. In this case, performance acceptability from a customer viewpoint often depends on the customer application, e.g. data transmission is much less tolerant of error than voice. In this instance, it is almost impossible to determine loss of revenue. Even if loss of revenue can be established, there remains the problem of apportionment of blame and recompense.

It would be reasonable to expect that all operators would have a minimum degree of integrity protection in their networks as a result of normal network-resilience measures. However, additional protection will need to be provided at the points of interconnection. These could, for example, be:

(1) detailed and unambiguous interconnect and feature standards. However, ensuring compatibility of implementation of standards

requires close co-operation between operators. The cost of creating standards by, for example ETSI, and its funding are well established. However, the costs of implementation and compatibility assessment can be very high and are bound to give rise to debate about how such costs should be funded;

(ii) mediation and policing devices. Their location (i.e. in one or both networks) and cost apportionment between operators will be a major issue;

(iii) conformance testing. This may require expensive captive models as well as internetwork trials, and it raises issues regarding when such testing should take place. This could be when new services are interconnected or even when major software upgrades take place in either network. Major issues include responsibility for conformance approval, self-certification, funding of test models and requirements for testing.

The commercial implications of an operator restricting access to its network in order to protect its integrity include:

(a) potential loss of revenue to the service provider or network operator denied access; this is difficult to independently quantify;

(b) the costs to customers who might be denied cheaper services; this is impossible to quantify;

(c) the reduction in market growth, innovation and development of new services resulting from the reduction of competition and lack of creative service providers. This is analagous to the creation of new applications for PCs. There would be an economic impact on the customer community; and

(d) conflict with the regulator whose policy is to open networks for interconnection to competing networks, e.g. the Open Network Provision (ONP) policy of the European Union and Open Network Architecture (ONA) policy in the USA.

12.3.5 Network and service management

Network management is the use of computer support systems to improve both the efficiency of maintenance and service provision processes, i.e. to reduce failure costs and to improve the performance of the network. Such systems are most effective when they interact with the network, each other and the operatives in a coherent fashion [3]. Hence much effort is being undertaken by the standards-creation bodies (ISO, ITU-T, ETSI etc.) to develop appropriate interworking standards under an architecture known as the Telecommunications Management Network (TMN). This has standard protocols and interfaces to allow maximum flexibility in interconnecting support systems and network elements in a multivendor environment. A well structured set of network-management support systems will markedly improve QoS in terms of time to provide service and repair. Additionally, it should detect problems before they affect customers.

No matter how good network management is, the impact on QoS can be reduced by a poor interface with the customer. While network management ensures that the operation of the network is optimised to the desired performance levels, service management should optimise services to the best customer QoS. Service management can be defined as the co-ordinated management of a portfolio of services which gives customers an integrated interface to the group of services that they use, and gives the network provider a single integrated view of its customers. It includes customer reception, to present a user-friendly single point for customers to interface with the network provider, hence making a significant contribution to the QoS perceived by customers.

While the development of network- and service-management support systems is very expensive, they make a significant reduction in prevention and failure costs as well as improving QoS.

12.3.6 Organisational costs

The way in which the field force is organised and directed can have a significant impact on both prevention and failure costs. A modern network of processor-controlled digital exchanges and transmission systems with remote monitoring and control features requires centralised control of operations staff via a minimum number of operations centres [5]. Such an organisation could contain a small number of network operations units (NOUs), the major control centres which initiate and control all activities. Also a central operations unit (COU), which provides top-level support, is the centre at which single tasks for the whole network are carried out and generally includes the national network-management centre which provides overall network oversight for network management, major network rearrangements and overall performance management. Network field units (NFUs) control field staff who carry out work scheduled by the NOU. Close links are provided to the 'front-office' customer-service units.

The development of work-management support systems can do much to reduce prevention and failure costs by increasing the productivity of those engaged in volume activities of provision and repair; primarily for customer-premises equipment (CPE) and in the local loop. Savings can be expected to accrue from more efficient day fill, skill matching, reduction in travelling time and distance, better communications, reduced control requirements, avoidance of work duplication, better jeopardy management and reduced effort in collecting and analysing statistics. Additionally, it can be expected that improvements to QoS would result from reductions in time to provide and repair, together with better control of appointments and quicker reaction to changing circumstances. The broad objectives of work management are to get the right person to the right place at the right time with the right stores and information; and know where operations people are, what they are doing, when they will finish and what they will do next.

12.3.7 Process-related costs

A process can be defined as the logical organisation of people, materials and procedures into activities designed to produce a specific result. Individual activities within a process are termed components and are normally linked together by exchanges of information or materials. QoS processes generally begin with a customer request and should end with a satisfied customer. However, for telecommunications, such processes are often complex, with a large number of components for a range of services and product lines modified by customer type and spanning a number of business organisational units. The opportunities for failures which affect QoS are therefore many. Process management is the control and systematic analysis and, if necessary, redesign of processes in order to maximise their efficiency and effectiveness [6].

Processes may be analysed in terms of the cost of each component, the time taken to execute it and, if possible, its quality. Where a process spans a number of business units, service level agreements (SLAs) are often negotiated between component owners at either side of the interfaces to record the QoS required across the interface to give the end-to-end target for the overall process.

12.3.8 Publicity and public relations

Good advertising and public relations often provide an effective way to improve customers' perception of QoS, which may not be directly linked to the service offered. However, national publicity is expensive, particularly if television is used.

12.4 Failure costs

12.4.1 Definition

Failure costs are those costs incurred as a result of unacceptable QoS as perceived by customers and the cost of rectifying failures in the network.

12.4.2 Network-related failure costs

Network-related failure costs are a large proportion of the cost of maintenance. In a modern network, a high proportion of these costs fall within the local-loop and CPE sectors. The minimisation of failure costs requires an understanding of the cost elements and their drivers, i.e. those aspects which influence them. Although the manpower element for operating the network is diminishing with the penetration of modern equipment and the introduction of computer support systems, pay is still a substantial element of operating costs. Time can be classified as effective and ineffective. The former can be measured and analysed to determine areas for productivity improvement, but the latter, covering such things as sick and annual leave, training and travelling are 'softer' and difficult to understand and analyse. They are often influenced by organisational structure, leadership and morale.

Many of the cost drivers are themselves dependent on other influences. For example, QoS targets drive capital and operating costs and are significantly affected by the regulator; they also determine market share which affects network growth and hence investment. There are also interactions between other network costs and drivers. For example, capital costs increase assets which incur additional depreciation charges and so drive up operating costs. However, if the investment is in modernisation then maintenance costs are reduced. Therefore, the full picture of costs, drivers and the interactions is extremely difficult to comprehend but it is an essential element of managing a telecommunications business. An example of the interactive nature of a cost driver is shown in Figure 12.5.

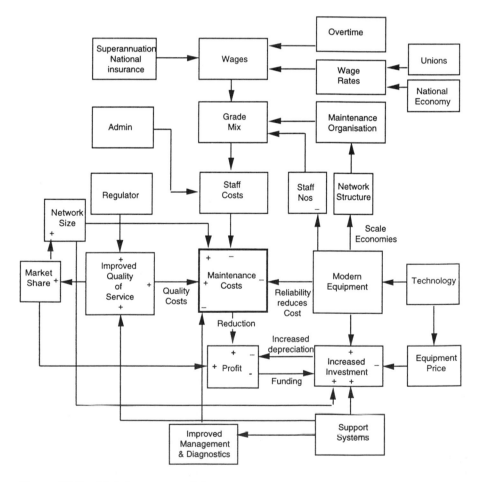

Figure 12.5   Maintenance-cost drivers

Failure costs can be minimised by a systematic analysis of failure costs to identify the drivers and thence determine the means of reducing costs. An example is given in Figure 12.6, where analysis of the operating costs has shown the local loop to be a high cost area. Examination of the loop costs indicates that maintenance (failure) costs dominate, so these are analysed to identify areas for improvement. The results indicate that, although primary cross-connect points (PCPs) are relatively few, compared with cables and DPs, they have a high fault rate. Further investigation reveals that failure costs for PCPs are driven by their design, poor workmanship when rearrangements are carried out and the number of times they are visited. Potential means of reducing the high fault rate are to refurbish the PCPs, replace them with a better design or to lock them and control access strictly.

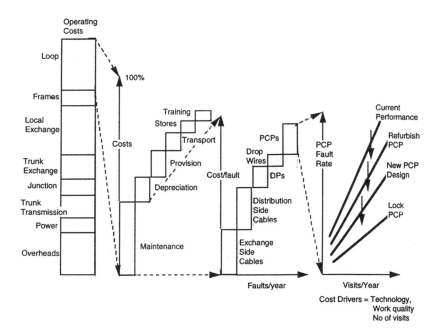

Figure 12.6   Investigation of failure costs

12.4.3 Network traffic management

A major cause of poor QoS from networks is call congestion caused by increasingly volatile traffic. This derives from the increasing usage of telephones, e.g. on Christmas Day, Mothers Day etc. and media-driven events which cause short-duration, focused surges of traffic to, for instance, TV and radio phone-in programmes and advertisements. Also, failure of increasingly larger modules of capacity, with the increases in the size of exchanges and transmission systems, has a great impact on traffic. Such heavy congestion cannot be overcome by traffic-routeing strategies such as automatic alternative routeing (AAR), and the resulting call re-attempts exacerbate the problem. This type of congestion may be classified as general overloads affecting the entire network, local overloads affecting part of the network, and focused overloads where the calling rate to a specific destination is particularly high.

The effect of these overloads can be reduced by network traffic management which monitors certain network-performance parameters from each exchange at regular, say 5 min, intervals. When they reach a prescribed threshold, build up of congestion is indicated. In such circumstances, traffic is managed by the use of expansive controls and protective controls. The former typically reroute traffic over parts of the network where spare capacity exists, to avoid congested areas. Protective controls are designed to prevent otherwise-ineffective calls to focused destinations from entering the network in order to allow other calls to succeed. The usual protective control is call gapping at the originating source; only 1 in n calls or 1 in T seconds is allowed so as to mitigate the effect of focused overloads. It does not improve QoS to the focused destination but improves it to other destinations.

The cost of network traffic management is high. However, it improves not only QoS but also the throughput of calls, thereby increasing revenues.

12.4.4 Revenue rebates

In a competitive environment, where the customer has a choice of supplier, network operators are often obliged to offer contractually binding SLAs which guarantee specific levels of service. For example, a standard contract for residential customers may specify a five-day time to provide and same-day repair; business customers may be offered superior QoS objectives. For any such contract, there is usually a penalty in terms of a revenue rebate for non-compliance; poor QoS can therefore result in a substantial loss of revenue.

In many countries, litigation is becoming increasingly commonplace and privatised network providers no longer have the protection of public ownership. Business customers which increasingly rely on good communications may well successfully sue for substantial damages if communications outages cause loss of income.

In an environment with contractual service targets against which failure penalties are paid, work-management processes and support systems must monitor tasks closely to ensure timely completion. Imminent and missed service targets must be selected automatically and referred to field managers for corrective action; this is known as jeopardy management.

12.4.5 Product launch

In a competitive environment, getting new products and services to market early is important in securing a high market share. However, poor product-launch processes result in 'teething troubles' and poor QoS to early customers and can cause loss of confidence in the service and poor market share. Failure costs are therefore high, in both loss of revenue and retrospective action to correct problems.

The launch of a new telecommunications product or service is a complex exercise which can involve a substantial multidiscipline team working to tight target dates. The core product team will often consist of a product manager who is responsible for overall product profitability, a marketing manager who represents customer requirements and a finance manager who produces the financial data associated with the product and who provides budgetary support. Expert support is provided by research and development, network planning and operations, customer-service operations, training, equipment procurement, logistics, commercial contracts, billing systems, tariffing etc. Good project management is therefore a key ingredient of the product-launch process. Such a project plan often uses critical-path techniques to sequence the activities, timescales, interactions, responsibilities and key milestones.

A feature-rich service realisation will provide good differentiation from competitors' services and could be more attractive to customers. This gives a high market penetration and share, and hence substantial revenue. However, such an option would require a high-functionality network and the complexity would incur risks in delivering to time and on budget, with high failure risk and cost. High-functionality services launched late may well lose market share to an earlier but simpler offering from a competitor. A low-

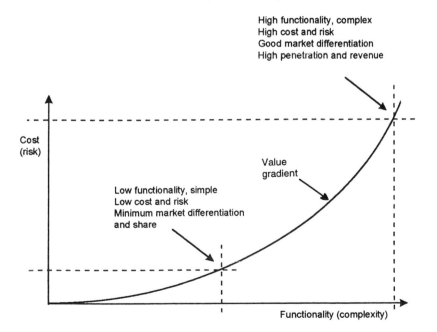

Figure 12.7   Risk assessment of network options

functionality option will generally be of lower risk but less competitive in the market. A compromise may be to launch early with a simple solution to capture the market, and then migrate the customer up the value gradient as shown in Figure 12.7. However, this requires a flexible network realisation to allow this to be done in a cost-effective manner.

For advanced services. e.g. intelligent-network services, the network-control complexity can give rise to a feature interaction, whereby new services interact with existing services in an undesirable and unpredictable manner. This can require high prevention costs when developing new services, to identify such problems by modelling using formal methods such as SDL (specification and description language) [7].

12.5 Appraisal costs

Appraisal costs represent the cost of measuring QoS and network perform-ance. The measurements range from, for example, customer-satisfaction surveys to the sampling of 1 in n live calls by exchange software and the processing of the samples to provide meaningful network-performance statistics [8].

12.6 Benefits

12.6.1 General

In a similar fashion to costs, the benefits of QoS are difficult to quantify. In a competitive environment, it is difficult for the incumbent network operator to compete on price for plain old telephony service (POTS), its main source of revenue, because of the universal-service obligation to serve all customers, even those which are unprofitable, at a fair price. For other services, the regulator will often insist that tariffs are cost based to prevent unfair competition by predatory pricing as a result of cross-subsidisation. Such operators often carry high overhead costs as a result of their historical monopoly culture. Therefore, in order to retain market share, the network operator is forced to differentiate itself from its competitors by offering a superior QoS and a more extensive portfolio of services.

12.6.2 Network modernisation

Good QoS and new services are difficult to achieve when a network operator is modernising its network because of the legacy of obsolete plant. Its competitors can 'build from scratch' with modern plant. It is therefore necessary for the network operator to schedule carefully the order in which exchanges are modernised. Early priority must obviously be given to large towns with high business penetration. These are most at risk in a competitive environment, but they are also likely to generate most revenue from new services. It becomes more difficult to prioritise the residual majority of exchanges to ensure that the supply of digital exchanges is deployed in the most cost-effective manner, particularly when replacing exchanges.

The financial worth of modernisation can be assessed by trading off the capital required to modernise each exchange against the savings in operating costs and the incremental revenue from new services, usage stimulation from improved QoS and the at-risk revenue which is protected from competitors. There is evidence that improved QoS can stimulate usage. For example, modernisation by replacing analogue exchanges by digital reduces call-establishment time and the probability of congestion while also improving call clarity though the reduction of transmission loss and noise. This tends to encourage customers, particularly in the residential market sector, to make more calls. Conversely, in a partially modernised network, the variability of performance due to the mix of analogue and digital routings can give rise to increased complaints about QoS.

Clearly, the reduction in congestion due to, say, automatic-alternative-routing and dynamic-alternative-routing strategies together with network traffic management increases traffic throughput and hence revenue. Some calls lost due to congestion are eventually successfully repeated. The 'repeat-call' button on most modern telephones assists in this, although it can exacerbate the problems of heavy congestion by exponentially increasing call attempts.

12.6.3 Productivity

It is generally acknowledged that failure costs in telecommunications outweigh prevention costs by many times. It therefore follows that increasing prevention costs to reduce failure costs will result in overall productivity gains.

12.6.4 Share price

When network operators are privatised, their worth, at any time, is determined by the price of their shares which is, in turn, determined by their perceived value by the stock market. The prime purpose of the stock (or equity) market is to provide a structured environment within which investment capital is raised. Hence, all equities of an equivalent risk class are priced to offer the same expected return. Shares are purchased by investors who require either high dividends, i.e. 'income shares', or capital appreciation of the share price, i.e. 'growth shares'. There is no fundamental difference between the two approaches; each delivers an equal return to the investor.

It is therefore possible to express 'shareholder value' as the sum of the dividends D and stock appreciation SA, i.e. shareholder value = $D + SA$. Hence the most attractive investments are those with high dividends and stock appreciation. To maintain a high and stable share price and hence company worth, a network provider needs to generate high profits from which to maintain good dividends or to be perceived by the market analysts and institutional investors as having good growth prospects and thus share appreciation.

Market share and company image in terms of the QoS delivered to customers are important influences on share price. A study by Aaker [9] found a strong relationship between changes in consumers' perceived quality and the stock performance of a corporation. It is suggested that major investors are directly influenced by changes in consumer goodwill for key brands, and they are sensitive to initiatives which have influenced, or may influence, consumer quality perceptions. Maintenance of shareholder value can be a significant driver of business policy of privatised network operators. If they do not exhibit growth potential then, to maintain share price, they may need to trade off against poor stock appreciation by paying high dividends which reduces the proportion of profit available for network investment.

12.6.5 Market share

It is generally accepted that customers make purchases on the basis of 'value for money' (VFM), where VFM is a function of perceived quality and perceived cost. Furthermore, as long ago as 1944, Scitovszky [10] argued that people judge quality by price on the basis that the forces of supply and demand would lead to a natural ordering of competing products on a price scale with a strong relationship between price and quality. Also, assuming that good quality costs more to produce, there would be a relationship between cost and quality and hence between cost and price if sellers set price in terms of cost plus profit margin. Much work has been done in an attempt to justify this, but generally in the context of the retail market.

Hill [11] argues that, as the level of quality required by customers increases, the value of a product also increases, as does the price. But the rate of increase in value decreases with the increase in quality level, while the rate of increase of cost, and hence price, will accelerate in the same circumstances. Hence, the best implied quality level for the customer is where the value/cost relationship has the maximum gap, as illustrated in Figure 12.8.

Figure 12.8   Price/value relationship

It is not clear what the price/quality relationship is in the relatively immature competitive telecommunications market. Generally, the previous monopoly supplier would be perceived as overcharging with poor QoS, and such an image is difficult to dispel. Furthermore, telecommunications pricing is complex in relation to normal purchases made by customers, i.e. a basic call-charge structure based on distance, duration, time of day or day of week, with possible discounts and price promotions plus rental and other services charges. It is therefore not surprising that customers' perception of price is normally poor and that they generally grossly overestimate call prices. This affects not only market share but also market size as witnessed by the poor utilisation of the local loop, i.e. of the order of 4–8 minutes per day on average for residential customers. This has led to advertising campaigns which are aimed at increasing the awareness of call prices, for example by comparing them with common retail products. Simplicity of pricing may well improve VFM perception.

Price elasticity and quality elasticity do exist but they tend to vary according to market segment[†] as illustrated in Figure 12.9.

It is likely that the market segment which is most price sensitive is low-calling-rate residential customers; this is normally the loss-making segment which the universal service obligation forces network operators to serve. This segment is, however, least sensitive to quality. On the other hand, businesses which are

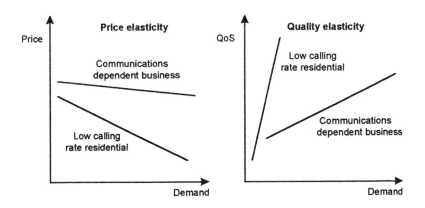

Figure 12.9 Price and quality elasticity

† Segmentation is the process of identifying groups of customers with sufficient common characteristics to make it possible to determine the market needs that each group requires.

dependent on good communications, e.g. the finance segment, tend to be highly sensitive to quality but less concerned with price. There are, of course, businesses which have a large communications spend but are also very price sensitive.

12.6.6 Brand

In a competitive market brand is an important differentiator. This is particularly important for the established telecommunications operator which finds it difficult to compete on price because of its large overhead-cost legacy from its monopoly era. According to Majaro [12], a strong brand brings reassurance to customers by providing a perception of permanence and quality. Not only has brand a major influence on market share, but any damage to brand image causes a slide towards the commodity end of the market with an inevitable reduction in price and hence revenue, as shown in Figure 12.10.

The perception of brand quality has a major impact on sales according to 'total research' which has been running the EquiTrend consumer survey since 1990. The central measure of EquiTrend is consumer perception of brand quality on a 0–10 scale. It is estimated that one point gain on the perceived quality scale can result in a 30% increase in sales [13].

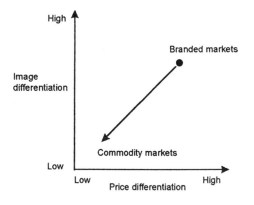

Figure 12.10   Brand behaviour

12.7 QoS model

A possible but subjective approach to evaluating the economics of QoS is illustrated in Figure 12.11. It assumes that the three main drivers of customer satisfaction are QoS, value for money and brand image. Thus, if two are held constant and the third varied, it would, theoretically, be possible using a survey questionnaire, to measure the change in customer satisfaction. If the thresholds of customer satisfaction at which various proportions of customers might defect were known, then it would be possible to calculate the loss of revenue.

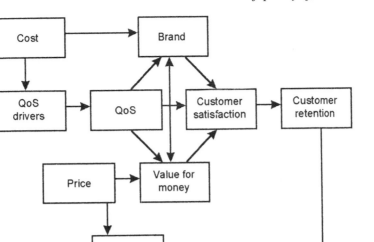

Figure 12.11   Customer-satisfaction model

If it were assumed that perceived QoS was the same as that achieved, then it might be possible to change a QoS parameter, e.g. time to repair, measure the cost and determine the likely retention of customers and hence revenue. Likewise, the effect of a price change and an advertising campaign could, theoretically, be evaluated. The effect of changing the variable could, perhaps, be measured by suitable questionnaires. However, there are many aspects which influence customer satisfaction, and which vary according to the market segment; isolating the three key drivers could be most difficult.

12.8 Conclusions

This chapter has dealt with the economics of QoS in a qualitative manner, since it is not possible to establish direct relationships between all of the cost and benefit elements. Nevertheless, these elements are large in financial terms and can have a major impact on network operator profitability. It is therefore important to understand the economic dimensions of quality and their drivers in order to have an informed view of the financial consequences of QoS initiatives or opportunities.

12.9 References

1 PENCE, J. L.: 'Is zero defects economical', Proceedings of *IEEE Globecom*, 1985, paper 5.3.1
2 PRIOR, J., and CHAPLIN, K.: 'The St. Albans study', *Bri. Telecom. Eng.*,1995, **14**, (1), pp.46–49

3   FURLEY, N.: 'The BT operational support systems architecture framework', *Bri. Telecom. Eng.*, 1996, **15**, (2), pp.114–121
4   Network Management Forum, 1994
5   MILWAY, N., and WRIGHT, B.: 'NAIP – the realisation of a network vision', *Bri. Telecom. Eng.*, 1995, **13**, (4), pp.268–273
6   FINEMAN, L.: 'Process re-engineering: measures and analysis in BT', *Bri. Telecom. Eng.*, 1996, **15**, (1), pp.4–12
7   WOOLLARD, K.: 'What's IN a model? modelling IN services using formal methods', *BT Tech. J.*, 1995, **13**, (2)
8   HAND, D., and ROGERS, D.: 'Network measurement and performance', *Bri. Telecom. Eng.*, 1995 **14**, (1), pp.5–11
9   AAKER, D.: 'The financial information content of perceived quality', *J. Marketing Res.*, May 1994
10   SCITOVSZKY, T.: 'Some consequences of the habit of judging quality by price', *Rev. Economic Studies*, 1944, Winter, pp.100–105
11   HILL, T.: 'Production/operations management' (Prentice Hall, 1983), pp. 275
12   MAJARO, S.: 'The essence of marketing' (Prentice Hall, 1993) pp.87–89
13   FOX, H. L.: 'As you like it.' Marketing Focus, 1995

Chapter 13

Role of standards

13.1 Introduction

The principal global standardisation body for telecommunications, the International Telecommunication Union (ITU), started as the International Telegraph Union in Paris in 1865 and took the present name in 1932. Work on standards intensified after the Second World War. One of their main objectives was to control and specify transmission loss over the analogue network. The loudness of end-to-end conversations had to be within acceptable limits. Studies developed into more sophisticated areas, e.g. traffic theory (to control congestion), sidetone, echo, delay (especially important over satellite circuits), post-dialling delay, noise of various forms and many other topics. The principal forum for discussing the studies was the then International Telegraph and Telephone Consultative Committee (CCITT) which was part of ITU, now known as ITU-T (International Telecommunication Union – Telecommunications Standardisation Sector). The principal role of the ITU-T is to develop and publish 'Recommendations' which are popularly called 'standards'. These standards are intended to cover all facets of telecommunications between different networks interworking all over the world. The topics covered range from provision of service to billing, and cover the principal activities during the life cycle of a service. The focus is on international interworking of networks. In many cases, the national portion of a standard may be apportioned from the international end-to-end performance. In these cases, there is no need for a separate national body to re-interpret the ITU-T's international end-to-end performance into the national portion. However, where multiple operators are present in a country additional end-to-end performance standards are usually required.

QoS is one of the areas addressed by the standardisation bodies. Owing to the long history of monopoly of the service providers in most parts of the world, for most of the history of telecommunications, there is a perceptible instinct among the providers to be inward looking and focus on the quality needs from their own points of view. However, because of increased liberalisation, competition, advances in technology, new applications, sophisticated customer-premises equipment, globalisation of services and increased demands from users, there is increased pressure on the standardisation bodies to reflect the needs of the changing times.

Standardisation bodies have focused on solutions for network performance and interworking. The recognition of the distinction between QoS and network performance appear to be a recent phenomenon, though there still appears to be some confusion. As there is increased pressure from users to be more attentive to their needs, standards bodies have to respond to this.

In this chapter the role of the principal standardisation body, ITU-T, is examined, together with its relationship with other parts of the tele-communications industry and the challenges facing it today. The chapter also illustrates how it can respond to current needs by examining the standards available for basic telephony and suggests what further work would enhance the quality of this service, and finally it suggests how the ITU-T could enhance the industry by slightly altering its mission for the future. The role of other standardisation bodies is also very briefly mentioned.

13.2 Standards and QoS

13.2.1 General

The purpose of a standard is to provide adequate information to fulfil the stated purpose or scope. To illustrate, a standard on speech transmission might have the following essential information:

- a measure of level of speech signal at the receiving end (expressed as loudness rating);
- level of undesirable signals (e.g. noise) at the receiving end, broken down to types of noise;

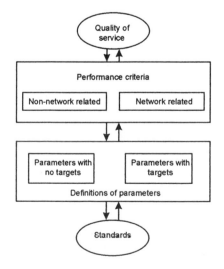

Figure 13.1   Schematic illustration of relationship between quality of service and standards

- a relationship between the percentage of users experiencing difficulty and combinations of various levels of the two above parameters.

The principal linkages in the relationship between standards and QoS are illustrated in Figure 13.1.

Standards may be classified into two categories:

- parameters defined but without target values;
- parameters defined but with target values.

An example of the first category is bids per circuit per hour (BCH). BCH is an indication of the average number of bids per circuit, in a specified time interval. It will therefore identify the demand and, when measured at each end of a both-way-operated circuit group, identify the direction of greater demand. It has no target but can be monitored to give an indication of network congestion, if a prescribed threshold (based on the size of the traffic route) is exceeded.

An example of the second category is the propagation delay. ITU-T *Recommendation G. 114* not only defines the propagation delay but also recommends international and national limits. If these are exceeded standards are not met.

The benefits of standards in relation to QoS in telecommunication may be discussed under the following headings:

- network providers;
- service providers;
- manufacturers;
- users/customers;
- regulators.

13.2.2 Standards and network providers

Network providers benefit from national and international standards on network technical performance. Standards could assist network providers to deploy systems faster and also free them from being locked into a single manufacturer. It is important that the standards for a new service, application or technology are published at the right time. Publish too soon and they are subject to costly amendments. Publish too late and they become ineffective.

Many major network providers have interfaces with networks of foreign service providers or network provider's administrations for the provision of international services. To specify the performances at interfaces and the projected end-to-end performance, it is necessary to an agree the performance provided by each national portion of the connection plus the international part. This task will be made much easier with internationally agreed performance standards.

Standards are necessary for the interworking among network providers within a country. In the UK there are, at present, two major network providers, BT and Cable and Wireless Communications (CWC), and many smaller providers. There are also an increasing number of cable-TV operators and cellular-radio operators have appeared on the scene. Standards could become a necessary means to ensure satisfactory end-to-end network performance for internetworked services.

Standards will be useful in specifying service level agreements (SLAs) with major customers. Targets could then be negotiated for improvement by mutual agreement between the network provider and the customer.

13.2.3 Standards and service providers

In cases where the network provider is not the service provider, nationally or internationally recognised standards will enable both parties to form the basis of contract for performance levels. For example, if a network provider provides the network and a service provider has responsibility for the provision of mobile phone service, then agreed performance-parameter definitions and standards will provide an easier and more manageable service agreement between the network provider and the service provider.

13.2.4 Standards and manufacturers

Recognised international standards enable manufacturers to develop equipment which can be sold throughout the world and will interwork with the equipment of other manufacturers. Well known examples of these are, Signalling System No. 7, and GSM for mobile telephony. In the absence of standards, these two areas of telecommunications would have not have penetrated and developed to the current extent. Manufacturers can assist in the identification of areas where such standards are to be developed and the practicality of embodying them in equipment design. Perhaps they could contribute by leading discussions in this area in the international standards forum.

13.2.5 Users and standards

Multinational companies may require standard definitions and values of performance to enable them to compare offerings from various service providers. If a multinational company provides private-network services to its sites in different parts of the world, standards will assist its estimation of end-to-end performance of QoS. Large customers may also wish to connect their private networks with public networks, and specifications of interface performances would facilitate easier SLA agreements with network providers. On a national level, standard definitions for performances will enable users to compare performance offerings of various service providers.

Service providers which do not co-operate in the provision of performance data to internationally agreed parameters, or which obstruct the creation of such parameters will be seen to be inward looking and protectionist and not really interested in catering for the benefits of the customer. Such attitudes are to be discouraged, not only from the customers' viewpoint but also for the good of the service provider, as in the long run customer confidence and loyalty will decline.

13.2.6 Standards and regulators

Regulators usually have a mandate to look after the customers' interests. To enable them to carry out this task, it is necessary for standardisation bodies to define clearly performance parameters for the principal services on a service-by-service basis and any performance standards that may apply. Regulators also have an important role to play in stimulating the creation of standards for the principal services and in ensuring that the service providers publish achieved performance results according to these defined standards. The role of regulators in the management of QoS is dealt in more detail in Chapter 15.

13.3 Standardisation bodies

13.3.1 General

The role of standardisation bodies may be appreciated by an understanding of the components in the end-to-end quality of a telecommunication service. Figure 13.2 illustrates these components.

The end-to-end quality of a service is dependent on the quality of the following components:

- quality of terminal equipment (also referred to as customer-premises equipment, CPE);

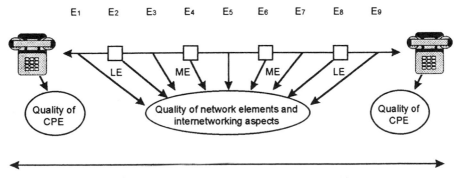

Figure 13.2   *Components of QoS on an end-to-end connection*
LE : local exchange
ME: main exchange
CPE: customer-premises equipment

- network quality comprising element performances (nodes and links E_1 to E_9, covered in Section 7.4.2 and Chapter 10);
- end-to-end service-surround and service-management issues (Chapter 9).

Examples of standards bodies may be conveniently classified in the following categories:

Approvals: BABT (in the UK).
National: BSI (in the UK); ANSI (in the USA).
Regional: ETSI.
International: IEC, ISO, ITU-T.

For a list of standardisation bodies see Macpherson [1].

Terminal-equipment approvals bodies
In countries where there is no liberalisation in the provision of terminal equipment, the end-to-end service provision and maintenance is probably under the responsibility of one dominant service provider which is also the network provider. In such cases the service provider ensures that the terminal equipment interworks with the network for optimum end-to-end quality.

In liberalised environments, where there are multiple suppliers of CPEs, it is necessary to have common standards of performance of terminal equipment and the interface at the access to the network. The specification for terminal equipment needs to address issues such as;

- electromagnetic compatibility;
- effective use of the radio frequency spectrum, where appropriate;
- protection of the network from harm by induced voltages from the terminal equipment;
- interworking aspects with the network; and
- performance of the terminal equipment for its principal function (e.g. voice quality for a telephone).

In the UK the terminal equipment for telecommunication services is approved by the British Approvals Board for Telecommunications (BABT). The majority of approved terminal equipment works satisfactorily when connected to the network in the UK. However, with the progress in harmonisation within European Union, the European Commission now requires all member countries to comply to the requirements in one of its directives, No. 91/263 [2]. This deals with the connection of terminal equipment to the network. The requirements have become less stringent and there is no guarantee that all terminal equipment which meet all requirements in the directive will function satisfactorily. The EC is currently developing a new directive 'Conformity assessment of connected tele-communication equipment in Europe' which will give manufacturers the opportunity to state compliance with EC requirements. This would enable national approvals bodies to relinquish some of their responsibilities in favour of the manufacturer.

The principal point to be noted is that an effective standardisation or approvals body is necessary to ensure optimum end-to-end quality of services. To ensure optimum quality of a service, the CPE quality should reflect the capabilities of the network and this, in turn, requires a recognised and authoritative body to lay down minimum standards of performance. In addition to approvals, the BABT is also involved in the comparative performance indicators and revenue-assurance schemes.

National bodies
Many countries have their own national standardisation bodies, such as the British Standards Institution in the UK. In the USA, due to the decentralisation, there are very many standardisation bodies. The American National Standards Institute (ANSI) has the *de facto* status of a national standardisation body. The national standardisation bodies are usually responsible for the introduction of standards for a range of topics ranging from agriculture to telecommunications and to X-rays. These bodies usually tailor standards from the 'parent' organisation, the International Standards Organisation, to suit local needs, where necessary, or publish the ISO version where no change is considered necessary. The national bodies may also produce specific national standards where necessary.

The regional standardisation bodies
Regional bodies are concerned with an acceptable level of quality and recommend standards to achieve this to a particular geographic area. The European Telecommunications Standards Institute (ETSI) is concerned with the development of standards for the geographical Europe. It liaises with the ITU-T where such standards may be of global interest. Where ITU-T is ahead in the development of Recommendations, ETSI manicures these, where necessary, to suit the needs of Europe.

International standardisation bodies
The International Eletrotechnical Commission (IEC) was formed in 1904 to facilitate the co-ordination of standards in eletrotechnology. Today its spheres of activities cover telecommunications, electronics and electrical and nuclear energy. It liaises with the ISO in the development of standards in telecommunications.

The International Standards Organisation (ISO) was founded in 1947 by 25 national standardisation bodies following a conference held in 1946 in London. Its objective is to encourage standards which will facilitate exchange of goods and services throughout the world. It works through the national standardising bodies such as the BSI in the UK. In areas close to telecommunications, such as information technology, it liaises with IEC, which also has a similar interest, in the development of standards to avoid duplication of effort. Such standards are published by ITU-T as one of its Recommendations with the logos of ISO and IEC to indicate their participation. The ISO publishes the same standard in its own standards-numbering scheme. The text is usually the same, with a very small number of the 160 standards developed by the ISO–IEC partnership having twin text. A list of these may be seen in the home page of the World Wide Web site of the ITU under study group 7. A recent work of ISO, a framework for QoS, is reviewed in Section 3.5.

The principal recognised body for telecommunications standards for the whole world is the Telecommunication Standardisation Sector of the ITU known more popularly in its abbreviated designation ITU-T. The ITU-T develops and publishes Recommendations [3]. Although it publishes Recommendations, these are widely considered as standards. In this book the term standard, as applied to ITU-T, is used in this context. In this chapter the principal emphasis is given to the work of ITU-T, due its global influence.

13.3.2 Telecommunications standards sector of ITU: ITU-T

Membership of ITU-T is open to all countries. They are represented by their principal telecommunication service provider, any recognised private operating agency (RPOA), government, regulators, users' representatives (user groups) and any scientific or industrial organisation with a legitimate interest in telecommunication services and networks.

The Recommendations developed by the ITU-T, are categorised under 26 headings, under each letter of the alphabet. The 'A series' covers 'Organisation of the work of the ITU-T' and 'Z series' covers 'Programming languages'. There are around 2500 ITU-T Recommendations taking up about 150 pages of A4 to list the titles. There are also Recommendations waiting to be formally approved, and a list of Recommendations which have been deleted. A list of these may be downloaded from the World Wide Web (www.itu.ch) or obtained from the ITU publishing department.

Recommendations are developed from the study of 'Questions' (i.e. areas for study) in 15 study groups, by some 4000 experts from around the world. Draft texts are written and presented for discussion at regular meetings of specialised groups; revised, and so on until the text is mature enough to be granted the status of a Recommendation. At this stage it is submitted for approval and subsequently published.

The lead study groups (which co-ordinate study within the ITU-T) in specific areas of study are:

SG 2   Service definition, numbering, routing and global mobility.
SG 4   TMN (Telecommunication Management Network)
SG 7   Open distributed processing (ODP), frame relay and for communication-system security
SG 8   Facsimile
SG 11 Intelligent network and FPLMTS (future public land-mobile telecommunications systems)
SG 12 Transmission performance
SG 13 General network aspects, global information infrastructures (GII) and broadband ISDN

SG 15 Access-network transport
SG 16 Multimedia services and systems.

Studies on QoS and related issues are carried out in study groups 2, 4, 7, 12 and 13. The lists of questions in these groups for the study period 1997-2000 are given in the Annex to this chapter (Section 13.7).

13.3.3 *Quality of service, network performance and ITU-T Recommendations*

Without the output from ITU-T, international telecommunications would not be possible. However, an examination of the Recommendations and the work in hand (see questions under various study groups in Tables 13.1–13.4) result in the following observations:

(a) Most of the QoS-related standards have been developed primarily to meet the technical-performance needs in the planning of a network and for interworking with other networks. However, some recommendations (notably the P series) were developed considering customers' responses to transmission quality; this includes a subjective element.

(b) QoS studies have mostly been stand-alone and do not form part of an architectural framework. An examination of the QoS-related recommendations indicates that many of the studies have a specific application and are not related to other recommendations. While these studies have been useful in their own right, the maximum benefit that might have accrued with the use of an architectural framework is missing. Absence of an architectural framework for the study of QoS has resulted in *ad hoc* and individual studies.

(c) There is lack of agreed definitions on QoS parameters, although such definitions would be of benefit to both customers and service providers. For example, for basic telephony there is an absence of performance-parameter definitions. This makes it impossible for the performances of service providers of one country to be compared with those of another country (see Chapter 16). A selected number of end-to-end QoS parameters, for the principal services, will not only benefit customers but will also be of benefit to the service providers.

(d) There is little or no mention of the cost of quality, nor any attempt to study in detail the concept of optimal level of quality. The optimal level of quality is governed by the customers' needs, service-providers' infrastructure, cost of quality components, and the price customers are willing to pay.

(e) There is no separate section (or series) for Recommendations on QoS. Those relating to QoS are distributed principally in the Recommendation series E, F, G, H, I, J, K, M, N, O, P, Q, R and V.

The classification of Recommendations has evolved over the years. Recommendations series are sometimes purely functional (e.g. D series on tariffs), sometimes part functional and part service (e.g. R series on telegraph transmission) and sometimes certain activities are grouped together (L series on construction, installation and protection of cable and other elements of outside plant). A series dedicated to QoS would, in the future, assist in focusing the issues related to quality and therefore should be considered by the ITU-T.

Perhaps the ITU-T could adapt to the increasing pressures and demands of the telecommunications industry by taking note of the above findings. Consultations with user groups, regulators and leading service providers would greatly contribute to a better understanding of the industry requirements.

13.4 Review of ITU-T Recommendations on QoS for basic telephony over PSTN

In this section the need for ITU-T Recommendations dealing with customer-related issues for basic telephony service over the public switched telephone network (PSTN) is examined. In the performance matrix shown in Figure 13.3, cells where Recommendations are required are shown. Any existing Recommendations, if customer related, are listed and their adequacy examined, and if further Recommendations or refinements are required these are indicated.

Service function \ Service quality criteria	Speed 1	Accuracy 2	Availability 3	Reliability 4	Security 5	Simplicity 6	Flexibility 7
Service management — sales & precontract activities 1	1	1	1	1	1	1	1
provision 2	2	1	1	1	1	1	1
alteration 3	1	1	1	1	1	1	1
service support 4	3	1	1	1	1	1	1
repair 5	2	1	1	1	1	1	1
cessation 6	1	1	1	1	1	1	1
Call technical quality — connection establishment 7	1	1	1	1	1	3	1
information transfer 8	1	3	1	1	1	1	1
connection release 9	1	4	1	1	1	1	1
Charging & billing 10	1	3	1	1	1	1	1
Network/service management by customer 11	1	1	1	1	1	1	1

Figure 13.3   Matrix indicating status of customer-related QoS Recommendations
1 no recommendations exist and none considered necessary
2 no recommendations exist, but some considered desirable
3 recommendations exist but refinements considered necessary
4 recommendations exist and considered adequate

Time for provision

No recommendations exist specifically dealing with the time for provision. This parameter is considered an essential quality-of-service parameter by most service providers, regulators and users. At present there is no internationally recognised definition for this parameter. Variables in the definition of this parameter are:

- calendar time or working time;
- if calendar time is chosen, the consideration of holidays;
- definition of the start and finish times for the provision, i.e. when the effective waiting time commences from contract and when the service is deemed provided to the customer; and
- whether the definition should be on a service-by-service basis or on a basket of services.

Desirable: An unambiguous and commonly agreed definition, developed by the ITU-T, will assist service providers and customers of different suppliers, nationally and internationally.

Time to resolve complaints

The principal Recommendation that exists for this parameter is E 420. However, this recommendation does not define the time for resolution of complaints. The issues to be addressed in the formulation of a definition are:

- identification of a complaint from an enquiry, clarification and other forms of queries from customer;
- the issues on the measurement of time as discussed above for the 'time for provision'.

Desirable: A universally agreed definition for the time for resolution of complaints could benefit customers and, possibly, service providers.

Time for repair

No recommendations exist specifically dealing with time for repair. Arguments for the time for provision, discussed above, apply to this parameter. The possible variables in the definition of this parameter are:

- the time when the fault is deemed to have taken place. Is it when it was reported by the user, or when identified by the service provider?
- when is the repair considered to be completed?

Desirable: A universally agreed definition or set of definitions developed by the ITU-T will assist both service providers and customers.

Simplicity in connection establishment

Recommendations specifying ring tones (indication when called person has been reached) exist. It would help customers if a uniform set of tones was specified for the following:

- indication of called customer's terminal engaged;
- indication of equipment engaged/busy/not available (network congestion).

Desirable: At present there is wide range of such indicators making it difficult for the customer calling an international number to identify which tone indicates which network state. Some standardisation could be useful.

Call quality (accuracy of information transfer, e.g. speech)
A measure of call quality for various types of calls may he helpful to customers. However, it must be added that, with digitalisation, the transmission quality of basic speech is considered less of a problem. Nevertheless, this parameter should still be studied, as it would be of interest in some parts of the world for some more years to come.

The principal Recommendations which cover transmission quality are:

ITU-T E.432	[4]
ITU-T E.855 Rev. 1	[5]
ITU-T G.101	[6]
ITU-T G.712	[7]
ITU-T P.11	[8]

The following are the degrading parameters contributing to lack of call quality (covered in P.11):

• loudness;
• circuit noise;
• sidetone;
• room noise;
• attenuation distortion;
• group delay distortion;
• absolute delay;
• talker echo;
• listener echo;
• nonlinear distortion;
• quantisation distortion;
• phase jitter;
• intelligible crosstalk.

A model to establish the relationship between customer opinion on a transmission path for a good signal and under various combinations of loudness, circuit noise, sidetone, room noise and attenuation distortion does exist. At present it is not possible to determine the combined effect of all parameters, as it becomes unwieldy and cumbersome to combine the effects of more than a few parameters.

Desirable: An indication of the level of call quality (expressed as the percentage of customers likely to find the circuit good for call quality) which users can expect for basic telephony for local, national and international calls under various circuit conditions.

Billing accuracy

The principal Recommendation on billing accuracy is E.433 [9]. This Recommendation quotes a maximum figure for the probability of under- and over-charging a customer. However this is not considered adequate. For sake of completeness, the magnitude of the maximum error should also be specified. Therefore an additional clause to this Recommendation stating the maximum permissible inaccuracy expressed in a suitable manner is considered desirable. For example, the maximum error should be expressed as not more than $x\%$ of the total bill or not more than $y\%$ of the correct charge to the call. The ITU-T could debate to determine the form of expression for the maximum permissible error.

Desirable: Refinement of the existing Recommendation to specify the magnitude of maximum error, in addition to the maximum frequency of error.

Examination of customer-related QoS Recommendations may be carried out on a service-by-service basis for their suitability. Refinements of existing Recommendations or addition of new ones should not introduce a conflict of interest with service providers. If such an exercise is carried out for all principal services, international comparisons of performance will be feasible. Additionally, customer appreciation of service quality will also be enhanced.

13.5 Future role

It is useful to paint a picture of the ideal of a standardisation body for the future. The standardisation body membership may indicated as shown in Figure 13.4.

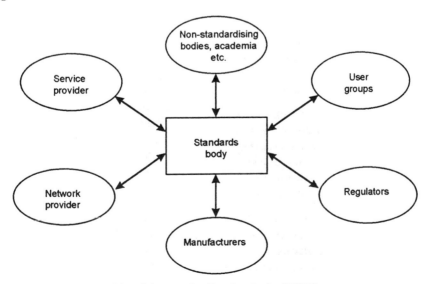

Figure 13.4   Membership of the standardisation body, ITU-T

Service provider
The service providers have been largely influential in the past in the initiation of topics to be studied in standardisation bodies. This situation is likely to continue but they should not have the principal voice in initiating and finalising standards. ITU-T members will have to take into account competition and the resulting need for a different emphasis in standards. For example, there is a need for standards on end-to-end performance parameters which will be of concern to customers. There could also be a need for more generic forms of standards, so that each service provider has the choice to provide a range of quality levels in a competitive environment. Standards need not be restrictive but should allow freedom to service providers to offer enhancement in quality offerings.

Network providers
The role of network providers in the standardisation bodies may remain unaltered in so far as their prime interest is concerned. Their primary role would be to ensure that standards are developed for interworking with other networks, both national and international, and to ensure compatibility of service usage. In the future, they will have to keep pace with new services and endeavour to initiate development of standards for publication at the right time. Multimedia services are currently being developed, and there is a need for standards on various issues. Network providers have a contribution to make in this area, not in isolation but in consultation with other bodies.

User groups
The user groups could influence the standardisation bodies in the choice and definitions of quality-of-service parameters for:

(i)   reporting achieved performance by the service providers and network providers;
(ii) specifying the offered QoS parameters.

The role of the user groups is discussed in more detail in Chapter 14.

Regulators
The regulator's role can be both reactive and proactive. Where quality has not been adequate, it can set national standards or suggest performance parameters on which to publish achieved results. With the introduction of new services, it could initiate development of new performance standards. The role of the regulator is discussed in more detail in Chapter 15.

Research organisations and academia and other organisations
Bodies such as the European Institute for Research and Strategic Studies in Telecommunications (EURESCOM), European Commission and academic organisations carry out studies on specific QoS topics. Their findings could be considered for their suitability for conversion into standards.

Manufacturers
Manufacturers can bring practical experience of the application of technology

and its resulting impact on the definition of the standards. Standards are economically applied if they can be easily embodied in telecommunications-systems development. In this respect, manufacturers usually have a significant contribution to make. In the future, with more applications and technology being enhanced, their contribution will be even more valuable.

The standardisation bodies

Standardisation bodies have to address the following issues:

(a) How do the ITU-T and other regional standards bodies decide when standards are to be developed? The computer industry and Internet are examples where lack of certain standards have helped development and expansion. Do standards inhibit innovation?

(b) Should standards be necessary in a competitive environment, or should the market forces be allowed to find their own equilibrium? When should there be intervention?

(c) With the convergence of telecommunications, information technology, information services, computers and broadcasting, how should the different standardising bodies of these industries work together?

(d) Should the standards be *de jure* or *de facto*? If there is room for both types of standards, what mechanism should be in place to channel the future standards into these categories? How much influence should market forces have on this aspect?

(e) Should there be more regional standards? If so should there be standards for interoperability?

(f) Where standards are required, how can these be developed in time for maximum use? How can this be facilitated? What are the catalysts?

(g) In the past, the ITU-T has concentrated mainly on technical performance of the network. In the future there will be additional requirements in the form of standards more oriented towards customers. How will the standardisation bodies adjust to these?

A conscious and reasoned effort must be made to choose a set of standards bodies to cater for most, if not all, known needs for telecommunications. The obvious choice is for one body to cover the whole world and regional bodies to specify requirements unique to a particular area, e.g. the European Commission for Europe. When regional standards bodies are recognised and granted authority and autonomy, the scope of their activities and their relationships with other similar bodies should be clearly specified. For example the European Commission should specify clearly the rationale and purpose of any directive to be issued on QoS to member countries. At present its policy is rather fluid. For basic telephony it suggests a set of parameters [10]. However, this is not mandatory and it leaves the member countries in a position to develop their own requirements over a period of time.

Where regional standardisation bodies exist, their functions should, where possible, be synchronised with those of the international body. They should specify changes in performance requirements from the global standards clearly and state how they are relevant to the particular region. Such an explanation will enable the network providers and service providers to be able, where necessary, to reconcile the differences in performance requirements.

Future standards work would benefit if these are based on an architectural framework for the study of QoS. At present there is no internationally agreed framework for studying all QoS parameters. However, a suitable framework has been proposed in an ETSI technical report [11]. This is dealt in more detail in Chapter 19. QoS standards will be most beneficial to all parties if they relate meaningfully to an internationally agreed framework.

The standards should cater for the viewpoints of all the principal parties associated with the telecommunications industry. These parties are the network provider, the service provider, the user/customer, the regulator and the manufacturer. The individual requirements are sometimes subtly different; if these differences are not identified and addressed in the standards, the opportunity for maximum benefit from the standards activity is lost.

The case for the inclusion of customers' QoS needs has been illustrated in Section 13.4. Examination of customers' needs whenever standards are developed should result in an increased benefit to customers.

Prognosis of future services and applications may be deduced from the progress in technology, developments in terminal equipment and new services offered by entrepreneurs. Study of these developments could lead to insight into areas where further standards may be required.

Standards, by their very nature, are time critical both in terms of time required for evolution and timing of availability for maximum benefit. For this reason alone, it is very desirable that the standards to be developed be prioritised according some credible criteria. Prioritisation is necessary because of limited resources.

Standardisation bodies have continually to address these and issues of the times in the future in positioning their roles in the telecommunications industry. Further discussion on the future of standards is in Section 18.7.

13.6 Conclusions

Standardisation bodies have played a useful part in the past. The changing requirements due to increased competition, formation of global alliances, convergence of telecommunications, broadcasting, computing and information industry will test the ingenuity of standardisation bodies. Their contribution in the future will also be tested severely by the increasing range of technical applications and the speed with which these applications are put into large-scale use. Correct timing of standards development will be a major issue. Other bodies, such as regulators, user groups and research organisations, could alleviate the difficulty of producing standards at this right time by making contributions to standardisation bodies.

13.7 Annexes

Table 13.1   List of questions to be studied by ITU-T Study Group 2 during the 1997-2000 study period

Question	Title
1/2	Applications of numbering and addressing plans for fixed and mobile services
2/2	Routing and interworking plans for fixed and mobile networks
3/2	Service quality of networks
4/2	Network management
5/2	Network-related QoS aspects of facsimile communication
6/2	Traffic engineering: performance objectives
7/2	Traffic engineering: measurement and modelling
8/2	Traffic engineering: dimensioning and control
9/2	Bureau services
10/2	Management and development of PSTN-based telecommunication services
11/2	New services and service enhancements brought about due to ISDN capabilities
12/2	New services for broadband ISDN (B-ISDN)
13/2	Mobile/personal telephone, telegraph, telematic, data, audio-visual and multimedia services
14/2	Service aspects of international multipoint communication via satellite
15/2	Universal Personal Telecommunication (UPT) service
16/2	Human-factors issues in telecommunications affecting multiple services or not related to specific services
17/2	Human-factors aspects of voice and nonvoice services using public terminals

Table 13.2   Questions to be studied by ITU-T Study Group 4 during the 1997-2000 study period

Question	Title
1/4	Terms and definitions
2/4	Designations in the international networks (circuits, group and line links, digital blocks, digital paths, data-transmission systems, digital blocks created between DCMCEs, virtual containers, multiplex sections etc., and related information
3/4	Maintenance of switched international circuits including telephone, ISDN and B-ISDN type circuits
4/4	Maintenance of mobile-telecommunications systems
5/4	Common-channel-signalling maintenance

6/4	Assessment of network performance and exchange of information for maintenance purposes
7/4	Fault, performance and configuration management of ISDNs and B-ISDNs
8/4	Maintenance of leased circuits and supporting transmission networks
9/4	Maintenance of digital transport networks
10/4	Test and measurement techniques and equipment
11/4	General aspects of test and measurement techniques and equipment
12/4	Quality assurance for TMN specifications
13/4	TMN principles, architecture and methodology
14/4	OSI system management
15/4	Requirements integration and management information/models for TMN interfaces
16/4	Requirements for the TMN F interface
17/4	Requirements for the TMN X interface
18/4	Network level management of transmission systems
19/4	Protocols to support operation, administration and maintenance at F, Q.3 and X interfaces
20/4	Protocols for the remote operation of management applications
21/4	Managed object definitions for management of telecommunication services, for network management and for network elements, based on TMN interfaces

Table 13.3   *Questions to be studied by ITU-T Study Group 7 during the 1997-2000 study period*

Question	Title
1/7	Technical characteristics, classes of service, facilities and categories of access for networks providing data communications
2/7	Network performance and QoS in data communication networks
3/7	Numbering plan for public data networks
4/7	Routing principles for public data networks
5/7	Principles of management for data networks and for the customer-network-management service
6/7	Interworking for networks providing data communication
7/7	DTE/DCE interface for packet and frame-mode DTEs
8/7	Non-native-mode terminal-access DTE/DCE interface procedures
9/7	Packet- and frame-mode signalling between public networks providing data communication
10/7	Lower-layer protocol and service mechanisms and features
11/7	Data compression
12/7	Network multicast

13/7	End-to-end multicast
14/7	Message-handling systems
15/7	Directory systems
16/7	Message-handling services
17/7	Directory services
18/7	X.400 and X.500 conformance testing
19/7	Open-systems architecture
20/7	Security services, mechanisms and protocols
21/7	Naming, addressing and registration
22/7	OSI application, presentation and session layers
23/7	Testing of data-communication protocols
24/7	Open distributed processing
25/7	Revision of recommendations

Table 13.4   *Questions to be studied by ITU-T Study Group 12 during the 1997-2000 study period*

Question	Title
1/12	Evolution of the programme of work
2/12	Definitions in the fields of telephonometry, speech signal processing, video signal processing, multimedia, terminal equipment and of characteristics of international connections and circuits
3/12	Radio-frequency effects on telecommunication voice terminals
4/12	Updating the 'Handbook on telephonometry'
5/12	Efficiency of devices for preventing the occurrence of excessive acoustic pressure by telephone receivers
6/12	Specification and test principles for hands-free terminals, acoustic echo cancellers and speech-enhancement devices
7/12	Analysis methods using complex measurement signals
8/12	General aspects in telephone elecroacoustic measurement
9/12	Speech-transmission characteristics and measurement methods for digital and handsfree terminals for both telephone band (300–3400 Hz) and wideband (50–7000 Hz)
10/12	Subjective methods for evaluating audiovisual quality in multimedia services
11/12	Objective methods for evaluating audiovisual quality in multimedia services
12/12	Cordless and mobile-terminal audio performance and testing requirements
13/12	Objective measurement of speech quality under conditions of nonlinear processing
14/12	Methods and tools for the subjective assessment of digital transmission systems
15/12	In-service nonintrusive assessment of voiceband-channel

	transmission performance
16/12	Transmission planning in the evolving mixed analogue/digital and ISDN networks
17/12	Noise aspects in evolving networks
18/12	Interconnection of private networks with the public ISDN/PSTN
19/12	Transmission-performance considerations for networks which are implemented using ATM technology
20/12	Analysis and extension of the E-model
21/12	Echo, transmission time and stability in multicarrier network environments

Table 13.5   Questions to be studied by ITU-T Study Group 13 during the 1997-2000 study period

Question	Title
1/13	New network capabilities for networks other than B-ISDN
2/13	Network capabilities required for the support of B-ISDN-based services
3/13	Network capabilities for interactive multimedia services
4/13	ATM layer
5/13	ATM adaptation layer
6/13	OAM and network management in B-ISDN
7/13	B-ISDN resource management
8/13	B-ISDN interworking
9/13	Interworking of 64 kbit/s ISDNs with other networks
10/13	ISDN frame mode bearer service (FMBS)
11/13	Enhancement and maintenance of ISDN layer 1 Recommendations
12/13	Access-network architecture principles and the interface functional characteristics
13/13	General performance issues (continuation of part of Q.16/13)
14/13	B-ISDN/ATM cell transfer performance (continuation of part of Q.16/13)
15/13	Availability performance (continuation of Q. 17/13)
16/13	Transmission-error performance (continuation of Q. 19/13)
17/13	Call-processing performance (continuation of Q.20/13)
18/13	Network synchronisation and time distribution performance (continuation of Q.21/13)
19/13	Transport-network architecture and interworking principles
20/13	Support of broadband connectionless data services on B-ISDN
21/13	General co-ordination of the network aspects for the support of interactive multimedia services
22/13	Use of the satellite transmission medium in the framework of the ISDN
23/13	General network studies

24/13	Global Information Infrastructure (GII)
25/13	GII principles and framework
26/13	Multimedia customer-access-layer requirements
27/13	Interworking between mobile and other networks
28/13	Vocabulary for general network aspects
29/13	Telecommunications architecture for an evolving environment

13.8 References

1 MACPHERSON, A.: 'International Telecommunication Standards Organisations' (Artech House, 1990)
2 Directive 91/263/EEC: 'The apportionment of the laws of Member States concerning telecommunications terminal equipment, including the mutual recognition of their conformity', April 1991
3 The full list of ITU-T Recommendations, Worldwide web: www.itu.ch
4 ITU-T E.432: 'Connection quality' (International Telecommunications Union, 1992)
5 ITU-T E. 855 Rev.1: 'Connection integrity objective for the international telephone service' (International Telecommunications Union, 1988)
6 ITU-T G.101: 'Transmission plan' (International Telecommunications Union, 1996)
7 ITU-T G.712: 'Transmission performance characteristics of pulse code modulation' (International Telecommunications Union, 1992)
8 ITU-T P.11: Effect of transmission impairments' (International Telecommunications Union, 1993)
9 ITU-T E.433: 'Billing integrity' (International Telecommunications Union, 1992)
10 Directive 95/62/EC of the European Parliament and of the Council of 13 December 1995
11 ETR 003: General aspects of QoS and network performance (QoS) and network performance (NP)' (European Telecommunications Standards Institute, October 1994)

Chapter 14
Role of user groups

14.1 Introduction

User groups exist to represent the users' interests in many industries and trades. User groups representing the users of telecommunications services have evolved, noticeably in the recent past. For other utilities, such as water, electricity and gas, which have reached maturity, the current principal user issues are related to distribution, cost of supply and perhaps also quality of the commodities. Telecommunications offers additional challenges to representatives of users on issues such as quality of services being developed, world-wide compatibility and interworking of services and greater interaction between the service providers and the terminal-equipment manufacturers for better use of network capabilities. With the prospect of rapid and continued growth in the range and use of telecommunications services and depth of technology, there is increased scope for user groups to influence various bodies involved in the provision of services. User groups could play an exciting role in the field of QoS, provided that they take the initiative to contribute effectively. They could contribute towards improved quality, coherence in service applications, user friendliness of the service–person interface, cost of service etc. These groups are in a unique position of understanding their member's needs on all aspects of telecommunication services.

Existing telecommunications user groups are involved in a wide range of activities. Typical activities are addressing issues on tariffs, provision of service at the required time, providing efficient customer care and QoS. These issues are usually taken up with the individual service providers. The more sophisticated user groups deal with wider issues, such as numbering, provision and coverage of new services (e.g. mobile) and regulatory matters. Some groups have taken the initiative on matters of QoS. However, it is the possible future role of these bodies which is given more attention in this chapter. Before these possibilities are explored, some of the principal concerns of three segments of user population are explored.

14.2 Principal issues of segments of users

14.2.1 General

For the purposes of analysis of the principal issues facing the users, the users may be divided into three broad segments: business users, residential users and those referred to as 'special-interest' groups. The latter comprise the disabled, the low-income group and anyone with special needs. The key concerns of these groups are first identified before ways are suggested in which user groups could contribute to addressing these issues.

14.2.2 Business users

In recent years, quality has become a major area of concern for all business users [1]. In the search for competitiveness, some manufacturers have cut back on factory testing and have sacrificed experienced customer-support staff, so that a growing problem among users is not just a lack of quality, but a lack of confidence in manufacturers and their products. To an extent, this concern has been projected at service providers as they, too, cut down on their staff strengths. At the same time, technological advances such as digitalisation and common-channel signalling concentrate traffic onto fewer high-volume routes, so that even a minor hardware failure or software bug can result in a widespread loss or degradation of services.

In the past, telecommunications, and later informatics, was an important, but subsidiary element of most businesses. However, their survival has come to rely more and more on information and the corporate network has become a major resource which ensures the flow, integrity and availability of all information. Users recognise that even small deteriorations in network quality can profoundly damage their businesses; hence, the need to monitor and assure the quality of the telecommunications services they use. In addition, their customers in turn demand high-quality services, and hence sophisticated internal monitoring is needed to guarantee their own performance levels.

Until recently, quality was regarded mainly as a technical issue, something to be 'built into' the design and manufacturing process, 'assured' by network and system design and dimensioning and 'maintained' by network monitoring and management systems and by competent and alert maintenance teams. The ITU and ISO were seen as the major guardians of telecommunications and informatics service quality, which tended to remain an esoteric affair and the domain of 'quality experts': the user 'called-up' the appropriate standards within the purchasing process and the vendor 'assured conformance' within those standards.

For many years, however, it has been recognised, that quality cannot be 'assured' by standards alone and exigent equipment users, such as military and government agencies, have always demanded vendor certification and regular inspection of the production process to create confidence.

Manufacturers have instituted 'total-quality' programmes and service providers have implemented sophisticated network-management systems to shorten outage times and improve service quality. User-oriented quality-assurance programmes have become marketing tools, widely advertised and exploited for product and vendor differentiation.

Business organisations may emphasise different quality indicators because they have different business interests or priorities, and some organisations may be more sensitive to certain quality degradations than others. Bit-error rates and outage rates may be tolerated by some users, but may be totally unacceptable to others, depending on their business application of communications.

Often users may have to interact with several different organisations to obtain the services they require, particularly, if the services cross national boundaries. The users' perception of quality then reflects the 'poorest performer' rather than the 'average performer', i.e. 'the weakest link in the chain'. It must be in everyone's interest to eradicate such problems, since perceived quality of telecommunications services is globally damaged by arbitrary and often avoidable regional differences.

Furthermore, the users' perception of quality often goes far beyond technical issues. Although the service provider and the service user are almost certainly organisations, people act as 'agents' for those organisations and interact with one another in the search for quality. Therefore, these 'agents' play roles, both within their own organisations and at service-provider/user meetings. While customers, which are bill-paying users, are particularly aware of costs and will tend to focus on individual cost–benefit relationships, providers will tend more to lean towards a general technically oriented solutions. Also, subtle interpersonal factors are called into the play, such as language and cultural differences, which participants may find hard to identify. Users often find that such problems are much more difficult to resolve than straightforward technical issues. The increasing inter-dependence and complex interactions encountered in international tele-communications make *ad hoc* solutions unworkable in the long term. Users need to do all they can to ensure the quality of the services and equipment they lease or purchase. The prime aim of quality monitoring by users nowadays is not to demonstrate lack of quality to a vendor, but to maintain internal and external, national and international performance of the services used.

In all cases, the measurement of QoS is of fundamental importance. However, it gives rise to serious problems, especially when the results need to be understood, agreed and interpreted by many different actors, including notably the service providers and the service users.

The user groups have expressed concerns on the dominance of major suppliers. These suppliers run the risk of exhibiting intellectual arrogance towards business customers arising from the technical knowhow they possess for the provision of new services.

14.2.3 Residential users

User groups have a significant role in representing the interests of residential users in a variety of contexts. Already apparent are national and international consultative forums on different aspects of policy, mostly associated with regulators.

Several factors combine to suggest that this role will grow:

- telecommunications is pervading everyday life, with society as a whole moving towards a new 'information society' in which telecommunications plays a vital part (see Chapter 18);
- the industry as a whole has a strong interest in ensuring that what it offers matches market needs;
- regulators are increasingly conscious of their duties towards consumers.

However, necessary resources for this important job may not yet be in place. At present, residential consumer groups specifically for telecommunications or information technology (IT) are not numerous. Usually the task falls to a generalist consumer body, which of course has to put forward consumer views on every service and good on the market, or to a telecommunications user group, and these tend to be dominated by business users.

Consumer representatives call for a dedicated independent group with adequate expertise and resources to advance the consumer interest in the information society at national level. The resources for such a body are unlikely to be found from membership subscriptions; if it is thought to be in the public interest, it will need public funding. This may be indirect industry funding, e.g. from licence fees or an industry levy.

The issues with which such a group will be concerned cannot be expected to fall neatly into any preconceived categories. Users do not perceive the same distinctions as an industry manager, for example between telecommunications carriage and content, or between quality and pricing issues. Examples of forward-looking consumer concerns found in a recent European survey include [2]:

- misuse of personal data;
- inadequate complaints and redress systems;
- insecure personal billing accounts;
- inadequate price indications;
- unclear contracts;
- unclear bills;
- unwanted calls;
- service content harmful or illegal.

At a more immediate level, users are already concerned from personal experience about system reliability, security and responsiveness.

User groups are most effective when they can combine the jobs of policy-level input with the day-to-day handling of actual users' problems. In this way, a feedback loop is formed whereby individual complaints can be seen as part of a larger picture, undesirable developments can be dealt with at an early stage and a positive contribution made to a more user-friendly future.

14.2.4 Special-interest users

In this section the situation of those special-interest groups whose requirements risk not being met by normal commercial processes in a competitive environment are examined. Groups in this category may include elderly people, those who are deaf, blind or otherwise disabled, and more broadly any needy group which it may not be commercially attractive to serve.

In most countries there is a social consensus that at least some of these groups ('the deserving poor') are entitled to the service provision and standards that they require, even if they cannot exercise commercial muscle to get them. Often this consensus is reflected in a requirement by the regulator to ensure that these groups are properly served. The regulator, the service provider or both must then find out what the groups do require.

Researching the requirements of these people directly may be particularly difficult because:

(i)   they may be hard to contact (a disproportionate number of them will not yet be on the phone);

(ii) they may not choose to identify themselves with the group in question – who wants to be thought of as old?

(iii) they may have little experience of telephone use;

(iv) some may have language or other communication difficulties;

(v)   the group in question may possess a membership organisation (e.g. Pensioner's Voice) which can be questioned. The officials of such a group are usually also members and are keen to articulate their perception of their member's interest. This can be very helpful. In interpreting the findings, however, it is important to remember that the membership is self selected (consisting of 'joiners') and may not be representative of the whole group. The officials display these same characteristics more markedly; their view of the member's interests may or may not be based on membership surveys or other direct evidence of member's views; and

(vi) the group may be helped by caring organisations, either official (e.g. local-authority social-services departments) or voluntary (Age Concern, Royal Institutions for the Blind and the Deaf). Similar remarks to those above apply. However, it is less likely that the representatives from such organisations will themselves be members of the group in question. It may be more likely that they will have available means of ascertaining the views of their 'clients', objectively.

A general public consultation can bring out the views of individuals who may, through their work or family, have much contact with the groups in question. As they are less likely to have a political agenda, the views of these individuals, taken together, may be especially valuable.

'Special' quality-of-service requirements fall into two main categories: different standards for the same aspects of service as concern others, and special aspects of service. Some examples for each category follow.

Special standards for same service aspects:

- extra fast repair, for those whose lives may depend critically on their telephone connection (e.g. people on life-support machines at home);
- operators trained to recognise callers with communication or intellectual difficulties and to handle their queries sympathetically, especially in emergencies;
- extra reliable payphones, for people who are dependent on these for all telephone service (e.g. aboriginal Australians).

Special service aspects:

- the availability of specially adapted equipment which is easy for those with limited sight or dexterity to use. 'Availability' here, importantly, encompasses awareness and affordability; special equipment is useless if those who need it do not know about it or cannot afford it;
- helpful treatment for people in payment difficulties, distinguishing the 'can't pays' from the 'won't pays';
- help with privacy concerns for those who feel vulnerable (c.g. those suffering form nuisance calls or who do not want to reveal their own number).

The customer's view of what constitutes QoS may not match that of the service provider or regulator – some requirements may be hard to fit in to the systematic matrix discussed elsewhere in this book.

However, it is worth pointing out that, for the majority in most special groups, the basic needs are exactly the same as everyone else's. For this reason also the group as a whole may choose not to express distinctive needs, preferring to share the general service quality.

There is another category of service requirement which may appear 'special' but is not, as it only emerges as a response to inferior service. This may best be explained by way of an example. Rural residents in both Australia and the UK are increasingly turning to teleworking, and may wish to use their telephone lines for data transmission. Some of these people find that, because of their distance from the exchange, the transmission quality is inadequate for the modem speed which they want to use. A demand for a guaranteed quality of data transmission thereby arises. Most urban residents are already getting adequate transmission and therefore do not perceive this as a requirement.

Once the areas have been found in which the distinctive QoS requirements of a special interest group lie, further steps are needed to:

- quantify targets for these requirements;
- where not all targets can be attained at the same time, assign priorities to them;

- report back on performance; and
- keep the set of requirements and target under review.

Again, both individuals in the special-interest groups and their representatives should be involved in these steps. For the quantification of requirements, objective research among the group in question is very useful. The representatives have a special role to play in agreeing priorities and reviewing overall views in the light of experience; random individuals could not be expected to have the necessary breadth of knowledge for this.

14.3 Possible contribution by user groups for the future

14.3.1 General

Owing to the specialised nature of needs of various segments of the telecommunication-user population, it would be advantageous if these needs could be identified separately. In certain countries, e.g. in the UK, many user groups exist. For example the National Health Service Telecommunications User Group (NHSTUG) deals with the requirements of the telecommunications service needs unique to hospitals throughout the UK. Other user groups are the UK ISDN User's Forum, Telecommunications Manager's Association (TMA), Telecommunications User's Association (TUA) and INTUG. Brief background notes on INTUG, TMA and TUA are given in an Annex at the end of this chapter. Further user-group details are given by Macpherson [3]. As the user groups have ready access to member's requirements on quality they are in a better position to ascertain an unbiased and unpressurised response from their members than is possible by service providers. This accessibility, when exploited by the user groups, could be used to the advantage of their own members, if necessary through the regulators. If the user groups take on such responsibility it could put them in a powerful negotiating position.

In addition to the user groups, there is a consumer interest group in the UK called the Consumers' Association which publishes customers' perceptions of various consumer goods on a monthly basis. Occasionally this includes telephone quality. Similar organisations exist in other countries, e.g. Consumentenbond in the Netherlands. There is also, in the UK, an industry-funded body called the Independent Committee for the Supervision of Standards of Telephone Information Services (ICSTIS). These bodies, collectively, can influence the service provider. The following are examples of how user groups could provide this support to their members.

14.3.2 Common set of definitions of parameters to express QoS

The lack of an agreed set of definitions for QoS is a disadvantage (discussed in Chapter 16). These disadvantages affect all users, but particularly business users. The QoS performance data published by the service providers are often couched in terms which mask the real needs of the customer. For example, a recent survey in UK showed that 95% of the respondents quoted a requirement for repair time of around 1.5 h or less. However, in the UK the published performance for repair is expressed, by most of the service providers, as $x\%$ of faults cleared in 'y' hours [4]. This does not tell the reader what proportion of the customer's repair needs are satisfied. It merely states how well the service provider's target has been met.

User groups could define their own sets of performance parameters. These could then be presented to standardising bodies for inclusion as parameters on which delivered performance ought to be published. The service providers could have their own sets of parameters for their own internal management purposes, in addition to these. The initiative to make this happen could come from the user groups.

14.3.3 Standardisation bodies and user groups

User groups are eligible for membership in international standardising forums such as the ITU and ETSI. While the majority of the membership is from the service and network providers, it is feasible for the user groups to initiate the adoption of standards which will be of benefit to them. Such standards ought to be complementary to those which represent the interests of the service and network providers. User groups could play a useful role in identifying new services, studying their implications and contributing towards setting quality standards in collaboration with other industry organisations. An example is the Multi Media Communications Forum (MMCF) in the USA, which has initiated the identification of many of the key issues (though not all) and is in the process of developing QoS standards for introduction into the ITU-T for the multimedia service (see Section 3.4). Such an initiative is highly desirable for any new service. The absence of such a contribution might result in a poorer set of QoS standards, dominated by the service provider's or the network provider's interests.

14.3.4 User groups and equipment manufacturers

Members of larger user groups, which are sometimes multinational companies (MNCs), are aware of the capabilities of equipment manufacturers. Some of these MNCs are in a unique position to influence the development of equipment for measurements on performance parameters. Even though the final development of performance-measuring devices would normally await the specification by standardising bodies and contract specification by service providers, nevertheless user groups have an influence in the development of measuring equipment. It is an opportunity for the user groups to take the initiative to influence an increasing number of service applications and the resulting increase in performance parameters.

14.3.5 User groups and regulators

One of the prime functions of a regulator is to ensure that the interests of users are met fairly by the service providers. To fulfil this function, user groups have a responsibility to 'educate' and advise the regulator on quality matters. In the European Union, the principal regulator is the European Commission, which usually seeks advice from user groups, service providers and independent consultants before a 'directive' is issued. Some of the directives which have a QoS content are listed in Chapter 16. Rather than wait for the regulator to seek out the opinions of the user groups, a forward thinking group could formulate policies for the best interests of their members for consideration by the regulator.

14.3.6 User groups and industry organisations

Various telecommunications-industry organisations exist which cover a range of activities. These range from education, seminars and conferences to studies into new service uses etc. The knowledge of users' quality needs, obtained by the user groups from their members, could be a useful input to these activities.

User groups could influence other industry organisations on matters related to QoS. Such organisations include the government ministries responsible for telecommunications, and also universities. These organisations could benefit from an understanding of the user needs in order to formulate policy and carry out fundamental research into quality.

14.3.7 Service providers, network providers and user groups

Perhaps the most obvious party to which the user could relate and provide support is the service provider. The user group could give the provider up-to-date information on what the members expect on QoS. User groups could advise their members on realistic levels of quality which could be achieved by the service providers. In many countries where user groups exist, dialogue between these and the service providers already take place; however, there appears to be scope for a greater involvement in the form of regular feedback to the providers on the levels of quality expected by the members, quality criteria to be defined for new services and selection of criteria for delivered quality on which statistics are regularly published. The relationship between service and network providers and user groups must be on a partnership basis for maximum benefit to all parties. To an extent, the service and network providers must concern themselves with the implications of commercially sensitive information imparted to user groups becoming common knowledge. However, in the current business climate it is considered beneficial to all parties to increase the level of co-operation among these parties to a mutually acceptable level.

14.3.8 User groups and quality-related tariffs

It is only right that the quality delivered by the service providers bears more than a resemblance to that offered or promised. It is also desirable to move towards an era where quality delivered will be linked to the tariff increases or tariff penalties imposed by the regulator. This point is discussed further in Chapter 15. The user groups could play a useful role in determining their members' opinions. Members have a first-hand knowledge of the revenue lost or at stake to them if the promised quality is not maintained by the service provider. Support may thus be provided to the regulator in establishing the formula for the tariff penalties as well as rewards based on the quality delivered.

14.4 Conclusions

With the increased sophistication in telecommunications, the role of the user groups could increase, from merely looking after the interests of their members to a supportive and partnership role to service providers, regulators and the industry in general in a more positive way. Absence of such an influence could result in the network provider and service provider imposing their concepts of quality on the customers with the possibility of not optimising customer needs with the network capabilities. Special-interest groups' requirements also need to be identified and channelled to the relevant bodies in order to do justice to the needs of members of such groups.

14.5 Annex: notes on three user groups

14.5.1 INTUG

Full title: International Telecommunications Users Group
Founded: 1974
Objectives: INTUG concerns itself with four major issues:

(i) monopoly authority and the rights of users;
(ii) free access to telecommunications networks;
(iii) freedom in user choice of equipment and services; and
(iv) constructive co-operation between public authorities and users.

Size: total spending power of members on telecommunications estimated to be several millions of pounds sterling. Membership of user groups through-out the world and large organisations.

14.5.2 TMA

Full title: Telecommunications Manager's Association
Founded : 1966
Main objectives:

(i)   education and information exchange;
(ii) liaison with government and other official bodies such as the Office of Telecommunications (OFTEL), Parliamentary Information Technology Committee, ETSI, OECD and ITU;
(iii) dialogue with suppliers; and
(iv) support for other interest and special focus groups.

Participated in OFTEL-led Comparable Performance Indicators Industry Forum, various joint BT and Mercury (now CWC) quality-improvement teams, membership surveys, co-ordinating consumer input to various official quality activities and is a member of the British Quality Association.

Size: Over 1200 members with an annual telecommunications budget (for equipment and services) of £3.6 billion. Membership is for individual Telecommunications Managers working in large organisations.

14.5.3 TUA

Full title: Telecommunications Users' Association
Founded: 1965
Objectives and principal activities: to represent its members' telecommunications interests and requirements in various bodies such as OFTEL, the European Commission (EC), suppliers and other relevant bodies.

Size: Membership is in excess of 1000. Estimated telecommunications expenditure is in excess of £1.5 billion a year. Membership is for organisations and not for individuals. Many multinationals are members.

14.6 References

1   LEE, A.: 'A user's perspective on QoS activities in world telecommunications', *Telektronnikk*, 1997, **93**, (1), pp 50–55
2   OVUM: 'The consumer in the information society'. Report for CEU/DG XXIV, June 1996, supplementary volumes
3   MACPHERSON, A.: 'International telecommunication standards organisations' (Artech, 1990)
4   'Comparable performance indicators – business customers'. OFTEL: published quarterly in the UK by the Office of Telecommunications

Chapter 15

Role of regulation

15.1 Introduction

Regulators in the telecommunications industry exist in many of the OECD†
countries and some countries outside it. The regulator's role is principally to
interpret the government's charter for the telecommunications industry. A
regulator normally has responsibility for one industry sector in one country.
The only exception is the regulatory role of the European Commission where
the regulatory aspects apply to any industry sector in all member countries of
the European Union. Increasingly, the role of quality in regulation is
becoming more important.

The parties affected by regulation are the users, service providers and the
regulators themselves. The management of quality takes on an additional
dimension: that of managing the implications of regulation. This chapter
looks at some regulatory considerations (some theoretical) of quality, the
evolution of quality regulation in the UK, some of the current issues on QoS
in the USA and a profile of the ideal regulator. Discussion on regulators, in
this chapter, has been confined to their role in QoS.

15.2 Regulatory considerations

15.2.1 Basis for regulation

For a theoretical analysis of regulation and quality, the reader is recommended to the work of Bowdery [1]. The following is a summary of part of
his work.

The basic case for the regulation of monopoly is that monopoly power
leads to the production of a lower level of output than the optimum. By
optimum level is meant that the cost of production is a minimum and therefore the price to the customer can also be a minimum, subject to a standard
mark-up in the profit. An allocatively efficient level of output is produced
when the marginal cost of the output equals its marginal benefit. Applied to
quality, the quality is allocatively efficient when its marginal cost equals its
marginal benefit. (In this chapter and elsewhere in this book the terms

†See footnote on page 284

allocatively efficient and optimal are used synonymously.) In an unregulated market the monopolistic organisation has the following options:

(a) The restriction of output below the optimum permits the firm to push up the effective price to the customers. This can also lead to higher dividends to the shareholders.

(b) The incentive to reduce costs and enhance profits by allowing quality to deteriorate provided that the demand for the firm's product is sufficiently quality inelastic that any loss of revenue following the quality deterioration is outweighed by the cost savings which accrue from it.

The above arguments can also be applied to quality. Optimum economic level of quality or the allocatively efficient level of quality can be offered to customers only when the cost of quality is minimum commensurate with demand and the resources at the disposal of the service provider. Normally, the regulator is much more concerned with competitive aspects of the telecommunications industry and the proper use of powers of the dominant company in the country. However, quality is increasingly becoming a key issue with the regulator. In the determination of optimum economic level of quality the following difficulties have to be addressed:

(i) Under normal circumstances in a single market, the benefit from the production of *an additional unit* of a homogeneous product (the marginal benefit) is only received by the marginal consumer (i.e. the consumer who buys the additional unit). However, an improvement in quality can usually be assumed to impact on *all units* of output. Thus the benefit from an enhancement in quality (the marginal benefit) includes the valuations of the quality enhancement by existing as well as those of any additional customers who may be persuaded to buy the product because of its improved quality. Therefore the simplistic approach of production levels of a homogenous product cannot be applied when quality is concerned. More work is needed to establish the relationship between quality, price and customer behaviour.

(ii) Changing tastes and technology make the optimal level of quality a moving target, even for a theoretical product with unidimensional quality for which there is undifferentiated demand. Changing technology impacts on both the nature of quality enhancements available to consumers and the marginal cost of achieving a given quality enhancement.

(iii) Telecommunications quality is not unidimensional. It is multi-dimensional and many of the economic models apply to unidimensional quality. This complicates analysis.

Despite the above difficulties the concept of optimal level of quality is important and it must be asked how far the various schemes available to the regulator will help to address these difficulties. There are basically three options available to the regulator. These are:

- publication of information on quality performance;
- service standards and compensation schemes;
- quality-sensitive price regulation.

These are examined in the following subsections.

15.2.2 Publication of information on quality performance

This is the most basic of regulatory instruments in the monitoring of quality from a service provider. It consists of a set of performance parameters on which the provider is expected to publish delivered quality. No targets of performance are specified for any parameter. In theory, the published levels of performance should indicate to the reader the performance attained by the provider. This form of quality data is cheap to publish and therefore easy to enforce by the regulator.

Some of the issues to be addressed in this form of monitoring quality are:

(a) Should the delivered quality data be audited by the regulator?

(b) There is no incentive for the secure and dominant service provider to improve the performance.

(c) The provider can, if necessary concentrate on the published parameters to show a good performance, to the detriment of other parameters.

(d) Data could be published selectively to give a misleading picture. For instance, the percentage of calls answered by the directory services within x seconds may be extremely high but not give any indication of the number of calls not answered due to network congestion or due to an engaged tone (i.e. not enough lines provided for this service).

(e) Publication of delivered data need not have any direct relationship with the allocatively efficient level of quality.

(f) No attempt is made to compensate aggrieved customers for poor quality.

15.2.3 Service standards and compensation schemes

In this form of regulation of quality, standards of performance are set for specified performance parameters and, if not met, providers are liable to pay compensation to customers. The compensation is aimed to address the balance of price–quality trade-off which had originally been offered to the customer. There is another rationale as well, and that is that the compensation scheme is expected to be an incentive for the provider to meet the individual standards set.

The following issues have to be addressed by the regulator in the formulation of service standards and compensations:

(a) In the assessment of trade-offs, the regulator has to make a judgement of the price–quality trade offs. With the knowledge available in this aspect of economic analysis, the consensus is that such a trade-off cannot be done accurately.

(b) In setting standards, the allocatively efficient level of service has to be determined. Any standards set without this consideration would not optimise the efficiency of the company.

(c) Setting of compensation should consider realistic disbenefits to the customer.

(d) Compensation levels should act as an incentive to the provider to meet levels of service promised.

(e) Choice of standards and parameters should reflect customer specific needs.

(f) Exclusion clauses in compensation schemes should reflect local business cultures. For example, the case both for and against exclusion of effects of weather conditions on performance in compensation could be argued.

(g) Universal service obligations.

15.2.4 Regulator-oriented quality-sensitive price cap

In this form of regulation, the provider's prices are subject to the levels of QoS delivered. The regulation takes the form of limiting permitted price increases to RPI$-x$, where, RPI is the retail price index and x is a factor determined by the regulator. In choosing the x factor the regulators take into consideration the capital costs, operating costs and an allowable rate of return on capital. The x factor is reviewed at intervals and the efficiency gains 'revealed' by the provider can be built into the x factor for the next period. The net effect is to improve the efficiency of the provider. Ideally, successive x factor regulation would drive the provider to provide a QoS that is the allocatively efficient level.

Bowdery goes on to state [1] that, in the absence of the ability to identify allocatively efficient quality levels, it might be assumed that quality levels are currently suboptimal. He offers the following arguments to support it:

(a) In conversation, a number of regulators and companies have stated that they suspect that current quality levels are suboptimal, although the degree of confidence with which this is suggested carries across both industries and quality dimensions.

(b) Technological developments and the dynamics of customer expectations might be expected to push optimal levels upwards.

(c) Rovizzi and Thompson [2] examined the belief that publicly owned utilities in the UK had 'gold-plated' and set quality above the allocatively efficient level. They looked at former state monopolies which were subsequently liberalised and concluded that it is difficult to believe that the results are consistent with the systematic overprovision of quality under traditional public ownership. Such findings provide some support for the view that, even some years after transfer to the private sector, quality levels still remain suboptimal.

Some of the issues to be addressed by the regulator in the study and treatment of x factor are:

(i)   how to close or curb the loophole whereby the provider can 'beat' the price cap by carrying out certain improvements only after the next review has been carried out. This would enable the provider just to meet the price-cap requirement of the previous period and not carry out further improvement until the next phase of x factor has been announced;

(ii) identification of the allocatively efficient level has been found to be fraught with difficulties and the identification of the optimal level of performance of the provider has been almost neglected. This is an area which requires further research, perhaps to be commissioned by the regulator;

(iii) how does the regulator deal with the situation whereby demand is quality insensitive? The provider has no incentive to improve quality;

(iv) the RPI$-x$ formula rewards the provider for past performance of quality. This does not deal with current performance levels and does not satisfactorily address the issue of future quality requirements. There is some evidence that regulators are beginning to address this issue. OFTEL, the UK regulator, has given some consideration to the expected quality of BT in the future, in the process of determining a price cap of RPI-4.5%, in 1997.

15.2.5 Customer-oriented quality-sensitive regulation

As the telecommunications industry becomes more sophisticated and competition increases both nationally and internationally, service providers will be subject to more pressures. Under these pressures, a spectrum of reactions could take place, varying from the emergence of world-class suppliers to the development of those that cut corners and offer the poorest service the market will stand. Such poor service may even extend to areas within a nation where there is little or no competition and therefore no incentive to provide a higher level of service. It will be therefore in the customers' interests for the regulator to ensure that poor QoS is penalised and provision of good quality is rewarded. It would be necessary for the regulator to establish a mechanism to identify good and poor service in an equitable manner.

One such method has been proposed by Lynch, Buzos and Berg [3]. The usefulness of this proposal, developed in a study in Florida, USA, has been checked and it has been found to be workable. Lynch *et al.* first describe the drawbacks where there is absence of reward or punishment in Florida (where there were 14 local telephone companies at the time of the study), develop a method of determining a formula for calculating the overall QoS of a telephone company and suggest a method for deciding the reward or punitive measures based on the delivered quality relative to the overall quality target.

In the scenario described, there were 38 QoS[†] parameters (grouped into 13 clusters and subclusters) in the state of Florida for the assessment of performance of a telephone company. Each parameter had a target value for performance. Similar setups existed in 30 other states of the USA. If the evaluation of a service provider is based on a pass/fail basis, the implications of various levels of superstandard and substandard performance are not given the consideration they need. Though companies were given performance standards, they are given little or no incentive to exceed those targets. Even the setting of standards was imperfect, based on a set of political and social forces. The setting of the standard appeared to be influenced by what was achievable by the least efficient. Even when standards were initially set based on marginal benefits and costs, technological advances may change the situation where the marginal benefits of exceeding the performance of certain dimensions would be much greater than the corresponding costs involved in implementing the change for the improved performance. In the system as it stood, there was no scope for the regulator to reward the service provider for such an initiative. Using the same argument, there was no incentive for companies to improve certain standards which fall below the required performance.

Another shortcoming in Florida was the lack of a procedure to arrive at an overall QoS figure based on the 38 performance parameters. In the absence of a formula, even the experts found it difficult to arrive intuitively at an overall figure. The consistency of arriving at a combined figure could decline with the increase in the number of parameters, according to various studies referenced in the original paper. In the absence of a credible overall quality figure, one option is to 'manage by exception', i.e. concentrate on the parameters on which companies have failed. This is also deemed unsatisfactory, since different companies fail on different parameters and therefore performance comparison becomes inconsistent. Studies also indicate that the decision makers become more *confident* of the accuracy of their evaluation with an increase of the number of dimensions of information, even if these increases in information are normatively irrelevant. Yet another difficulty in evaluation of overall quality is that the decision makers use less cognitive effort as the number of dimensions increases.

In the face of these difficulties, the only option remaining was to count the number of pass/fail parameters for different service providers. This has the weakness that parameters which are considered important may not be given as much weighting as they deserve and would be treated on a par with less important parameters. The paper illustrates a case of three companies: company A just meeting all standards, company B failing on one parameter and company C failing on two parameters. With the simplistic approach, company A would be deemed to be the best in terms of quality. However, company B was deemed to provide the best overall quality due to the lesser

---

[†]Since the work of Lynch *et al.* the number of parameters has been increased; the current list is given in Section 16.3.4.

importance of the parameter on which it failed and the improved performance on a parameter considered more pertinent to quality.

The authors, after describing the shortcomings in the method of regulators' judgements, propose the development of a formula to represent more realistically a quality index for which service providers may be rewarded or punished. The authors proposes a methodology for the determination of a representative index for the achieved or delivered QoS provider. Using techniques associated with the information-integration paradigm [4, 5, 6, 7], it is possible to determine a mathematical function relating the numerical value of technical measures to perceive overall quality:

$$Q = f(x_1, x_2, \ldots\ldots, x_n)$$

where Q is the index of quality, f a function, x the parameter and n the number of parameters.

The authors claim that this system of estimation of quality index has four main advantages. These are:

(a) The estimated model leads to greater consistency in the evaluation of quality, capturing reliable portions of their judgement policies, and eliminates the unreliable elements, as found by Bowman [8].

(b) Both political considerations and the psychological consequences of information overload are avoided. Thus the mathematical model can be a more valid tool for assessing overall quality than the unaided judgement of the regulator.

(c) This method leads to discussion where everyone can focus on dimensions of quality which are more or less important, rather than on the particular company being evaluated. This conclusion is supported by work by Edwards [9].

(d) It communicates clear and appropriate incentives to the regulated companies that should allow them to raise quality while reducing costs.

A detailed analysis of this work is not undertaken in this book. Readers are recommended to read the work of Lynch *et al.* [3]. According to a NRRI report [10] this type of weighted index has two limitations:

(i) The measurement requires that weights be assigned to each service attribute by the regulator. Lack of information on costs and customer preferences, which is necessary for establishing appropriate weights on each attribute, may lead to inefficiencies. In the work of Lynch *et al.*, a method to minimise this limitation is to use the expert opinion of the regulators.

(ii) The approach does not necessarily encompass all service attributes valued by customers but simply overlooks those unmeasurable by regulators. These excluded quality dimensions provide another potential source for inefficiency. Therefore the measurement proposal remains susceptible to some of the same informational problems and economic biases characterising the traditional forms of direct quality regulation.

The NRRI report then recommends further work [11] which appears to overcome the above limitations.

Despite the apparent complexity of this method of determining a 'performance index' to ascertain a fair assessment of a service provider's delivered quality, a feasibility study can be recommended for its implementation by a regulator. Such a study may perhaps be initiated through universities, in the form of academic theses.

15.2.6 Different price/quality combinations

Demand for telecommunications services could exhibit differentiated quality demands. In other words, different segments of the customer population could require different levels of quality, to suit their particular needs. This means that the service provider has to offer different price/quality combinations. The regulators favour these offerings. However, for the regulator to monitor whether the offerings are optimal, the following issues have to be addressed:

(i)   Are the different offerings having an unfavourable effect on the standard offerings or the customers who require standard offerings? How are the possible resource misallocations to be identified?
(ii) How does the regulator set standards for the different quality segments?
(iii) How does the differentiation take place? Where do each different segment's boundaries lie?
(iv) How does the regulator set compensation schemes for differentiated quality offerings?
(v)   Differentiated segments could also include the disabled with special quality requirements. The regulator has to identify the needs of such special segments of the customers, and address their quality needs to ensure that the provider does meet their quality requirements.

Notwithstanding the above issues to be addressed, this is an area, where, according to Bowdery [1], the regulators are enthusiastic for the providers of regulated industries to provide differentiated price/quality offerings.

15.3   Evolution of regulation of quality in the UK

Regulation of the telecommunications industry in the UK was born with the privatisation of BT in 1984. Under the 1984 Telecommunications Act [12], British Telecommunications became a public limited company (PLC). BT and Mercury (now CWC) were licensed under the 1984 Act to provide public-fixed-link telecommunications service. Apart from setting out BT's public-service obligations in areas such as emergency services, call boxes, rural services etc. the license said nothing explicitly about QoS. Indeed when BT's licence was being drafted, a condition designed to regulate QoS was strongly opposed by BT on grounds of the difficulty of defining QoS requirements both precisely and flexibly. BT's view prevailed. The nearest reference to quality was the statement to 'promote the interests of consumers, purchasers, and other users in the UK ... in respect of ... the quality and variety of tele-communication services' and 'to maintain and promote effective compe-tition' within the industry. The 1984 legislation did not give the Office of Telecommunications (OFTEL) specific powers or duties with respect to QoS.

The RPI – $x\%$ formula was reviewed a few times. It has been reaffirmed that quality must be seen as the 'other side of the coin' of price control – a quality decrease can be seen as a hidden price increase. The question has been raised of whether the price-control mechanism should be extended to deal explicitly with quality. For example, some sort of quality index could be devised which would have to rise at a prescribed rate, or else trigger price cuts. To date this approach has been rejected, although not conclusively. OFTEL has stated: 'if significant deterioration in service quality were to occur, this could be regarded as a reason for looking again at the level of the price cap' [13].

In 1986 a consultative document on QoS was published [14]. However, nothing much came out of this. BT, which had ceased publication of QoS statistics after privatisation on grounds of 'commercial sensitivity', recommenced publication of such data. That year also marked OFTEL's commencement of an annual report on quality of telecommunication services. OFTEL's findings indicated that BT's quality had worsened in areas of line repair, provision of new lines, communication quality, operator and directory-enquiry services and public call boxes. A survey by the NCC showed that BT's QoS was seen to be the 'worst' of the utilities [15].

In 1989 BT offered contractual agreement for payment for delays in fault repair and the provision of new lines. In the summer of 1991 targets for fault repairs were tightened. In 1992 the government passed the Competition and Services (Utilities) Act [16], known as the citizen's charter. Further quality regulation were proposed in this act. The power of OFTEL regarding quality was slightly increased, though worded in very vague terms. Also in 1992, in a consultative paper on the regulation of BT's prices, OFTEL explicitly discussed the possibility of using adjustments to the price-control formula to stimulate the production of allocatively efficient service levels. However, OFTEL admitted its inability to identify the marginal benefits and costs necessary for the determination of the allocative efficient level of quality. It was conceded that, to a degree, the incremental cost of improvement in quality could be determined. OFTEL also pointed out the problems to be faced due to the multidimensional nature of the QoS, the differentiated demand for quality dimensions and the related problems in constructing weighted quality indices. These proposals were not pursued much further.

An industry workshop forum was set up around October 1993, with OFTEL acting as a facilitator to arrive at a set of performance indicators in which service providers could report delivered performance. These parameters were measured so that different suppliers could produce comparable data. The first of these data were published by OFTEL in spring 1996, for the October – December 1995 quarter and continue to be published every quarter. These parameters will be supplemented later with technical parameters.

Evolution of quality regulation in the UK shows that most, if not all, steps taken have been reactive. There is little evidence of foreseeing problems associated with quality delivered to the customer and attempting to address these.

15.4 Regulatory issues in the USA

In a recent study by the National Regulatory Research Institute (NRRI) in the USA, the issues related to QoS in that country were addressed. The findings and the reasons for regulation are worth noting. The following is an extract from their report [10]. The NRRI looks at the voluntary industry controls, market controls and regulatory controls.

15.4.1 Industry controls and quality

Some of the most important decisions on telecommunications service quality, determinations that will affect consumers for years to come, are being made through processes which promote victory for the most powerful players, not necessarily the best or most economically efficient ideas. The process of identifying and agreeing on industry standards is complex, political and not necessarily internally democratic. Decisions on who gets what, when and how are constantly being made by participants with varying levels of power. These decision makers are not responsive in any direct way to the public. For good or ill, customer interest is assumed to be represented through company interests.

Voluntary standards organisations (VSOs) have been formed by users and producers in the telecommunications industry to debate and adopt standards. The process is cumbersome and slow. Most importantly, customers are notably absent from the discussions. Many decisions which affect the public switched network are being made outside the standardisation organisations, where protocols are developed on the basis of 'whoever thought of it first'. In neither case – the rigid, slow processes of VSOs nor chaotic development – are customers represented consistently and adequately. The public has a strong interest in the development of today's networks that ensures reliability, and is held hostage neither to weak links nor the subtle exercise of monopoly power which sets the parameters for millions of electronic transactions

15.4.2 Market controls on quality

Companies compete on the basis of quality as well as price, and economists tend to agree that increased competitiveness generally leads to increased experimentation with levels of quality. This benefits customers, which can choose the types and amount of quality they want at the prices offered. A protected monopolist lacks incentives to respond fully to the potentially wide range of customer preferences for quality. Thus, despite some potential drawbacks of competition, such as brand proliferation which may be used by an incumbent to attempt to block entry, customers are better serviced by competition than by monopoly.

Companies with monopoly power are likely not only to provide less variety in the services they offer, but to distort levels of quality and discriminate against low-end customers. Whether the monopolist provides lower or higher quality than demanded depends on whether the services are substitutes or complements. At a fixed output, when output and quality are demand substitutes, the monopolist selects a lower-than-optimal level of quality; when output and quality are demand complements, a higher-than-optimal level. The monopolist reduces quality for the customer at the low end of demand, not out of a direct desire to do harm but because this enables more customer surplus to be extracted from the high-end users. This is a particularly important point for public-utility regulators to remember: given the opportunity, the telecommunications firm which retains substantial market power will attempt to reduce quality for users of basic services in order to encourage to the purchase of better service by those able to afford it.

15.4.3 Economic and protective regulatory controls on quality

As the form and applicability of economic regulation changes, public utility commissions (in the USA state commissions) have been strengthening protective regulatory controls on quality and tying them more closely to economic regulation. Staff at 32 such commissions participating in an NRRI survey conducted in the spring and summer of 1995 reported many reasons for initiating or revising QoS standards in their states. The primary reasons were new technology and the actual or potential deterioration of service quality. The propensity of price regulation to encourage reductions in quality is a major concern among the commissions. Fourteen jurisdictions had tied their new or revised QoS standards to an alternative regulation plan. In some cases, a price cap formula included a service-quality factor.

Weighted indices of quality are being used in at least four states (see Section 15.2.5 for an example of a weighted index). An overall quality index is an improvement over traditional standard setting by making commission decision making easier once the index is developed and agreed, and allowing companies flexibility in how they meet service-quality requirements.

Commissions use several means of monitoring the QoS offered by jurisdictional telecommunications utilities. These include company reports, customer complaints, field investigations and customer surveys.

Whatever the form of regulation of quality, whether it is traditional standards setting, standards tied to price regulation or a weighted index, inadequate information can make supervision imperfect. Not all relevant service characteristics can easily be measured, and, even when they are, neither the costs of supplying quality nor the demand customers have for quality can be evaluated with any accuracy.

15.4.4 Advantages and disadvantages of market controls, industry controls and regulatory controls

The principal factors used in the evaluation of the three approaches to quality controls are:

(i)   meeting customer demand for quality;
(ii)   improving industry economic performance;
(iii)   adaptability to change and fostering of innovation;
(iv)   low administrative costs;
(v)   ability to meet industry demand for quality;
(vi)   achievement of equity objectives;
(vii)   economic development; and
(viii) ability to measure impacts.

Not every factor has equal weight. A thorough analysis would look at each of the six dimensions of QoS (availability, reliability, security, simplicity, flexibility/choice and assurance) individually for each of the eight factors in order to assess the cost–effectiveness of the three approaches.

Even a cursory look at the relative advantages and disadvantages of the control mechanisms leads to the conclusion that, compared with the other two, a market standard has impressive advantages. For the first four criteria listed above, effective competition (if it can be attained) is the preferred means of achieving quality. Administrative costs would be low to nonexistent. The ability of firms in competitive markets to align themselves with real customer preferences and in the process to maximise flexibility and choice is unsurpassed. Many telecommunications firms are already competing on the basis of quality. Bell Atlantic has promoted its reliability. Ameritech has run radio advertisements suggesting one-stop shopping for all the customers' telecommunications needs, an effort to compete on the basis of assurance. Articles in the trade press have emphasised the importance of companies' customer service [17–19]. The ability to innovate and adapt to changing conditions in the business environment is far superior in a market than under any kind of hierarchical control mechanism, whether imposed by industry or government. Industry economic productivity should improve, as budgets are appropriately revised and market-based investment decisions made. Some of the service-quality problems that regulated telecommunications companies have had may be due to inexperience in responding to the voice of the customer. As they gain familiarity with demand and marketing, companies may be better able to make business decisions which do not focus merely on cutting cost but on customer service as well.

Determining when market is sufficiently competitive so as no longer to need customer-protection standards is, of course, the key public-policy question. At a practical level, one test might be the number of customer complaints about telecommunications services. If they dropped substantially, then a competitive telecommunications market might (in the absence of other information) be assumed to exist. Developing and applying clear criteria to identify a competitive market will be essential to making correct public-policy decisions which affect the quality as well as the price of telecommunications services. One such set of criteria has been developed by Edwin Rosenberg of the NRRI staff [20]. The Telecommunications Act of 1996 includes a competitive checklist to guide judgements on when local markets are competitive. The feasibility of market controls depends on how well the market has developed and an accurate assessment of the degree of competition by government agencies, whether they be the state commissions, the Federal Communications Commission (FCC), or the Department of Justice.

Where competition does not yet exist, administrative costs are likely to be higher if regulatory rather than industry controls are imposed on quality, while adaptability to change and the ability to foster innovation may be lower when government intervenes rather than when industry regulates itself. Meeting customer demand for quality is likely to fall short under either industry or regulatory controls. The company with monopoly power will tend to undersupply quality when output and quality are demand substitutes, oversupply when they are complements and reduce basic service quality while introducing high-price service enhancements. Well designed regulatory programmes limit the ability of the monopolist to use these strategies, although experience shows that the result may be an oversupply of reliability and assurance and an undersupply of choice of services. Under monopoly conditions, improvements in industry performance are best achieved by coupling price regulation with QoS incentives.

Ability to meet industry demand for quality (the fifth factor to be considered in the truncated approximation of a cost–benefit analysis) could not be accomplished fully even if there were perfect competition, in so far as the technical needs of establishing and maintaining an intermeshed network are concerned. The incentive to establish and comply with standards comes from the need of the owners of telecommunications networks to send and receive the traffic carried by other networks. They can be expected to aim for high reliability. Although they have strong reservations about the process of technical standard setting, that process is moving swiftly and inexorably. It would be neither feasible nor desirable for state regulatory commissions to intervene actively in the process of setting technical standards. Observation of the standard-setting process by government agencies representing the public would be desirable. The role of commissions as mediators or arbitrators, provided by the federal telecommunication-reform legislation, makes sense where incumbent carriers attempt to leverage monopoly power to their advantage in setting and adhering to QoS standards for interconnection. In other words, where interconnectors are the customers and one provider still has monopoly power, government oversight of the service quality provided to them is justified as it is for pricing issues such as access charges.

Achievements of equity objectives, like universal service and the furthering of economic development, are the domain of government intervention rather than market or industry controls. The spectre of a country divided into information 'haves' and 'have-nots' might well come to pass without some government supervision. Although opening markets to competition is likely to lead to greater productivity and world-wide competitiveness for US companies, industry use of discount rates which emphasise short-term profits rather than long-term social goals can lead economic growth which is uneven.

Measurability of levels of quality achieved is the final factor to be considered in deciding which form of control is appropriate for assuring quality at levels that customers want. Without the ability to assess quality, it will not be possible to see whether public-policy objectives are being met. Nor will customers be able to compare quality choices systematically. Whether market structure is competitive or affected by monopoly power, industry will have little reason to collect and publish statistics on quality, and incumbent companies can use their brand names to hold onto customers. Professional quality analysts are not likely to spring up in the private sector to help customers evaluate quality offerings the way financial analysts help investors judge the value of financial instruments. Developing, supplying, and publishing measures of quality in telecommunications is best accomplished by governments. By having a public measure, one encourages new entrants because they can spend less capital on building a name and more on complying with the standards.

15.5 Future of regulation

15.5.1 Global regulation

Early telecommunication regulation was mainly confined to domestic boundaries because service providers provided international services on a correspondent basis with bilateral agreements on tariffs between countries. However, the recent emergence of regional regulators, the prime example being the EU, are increasingly influencing the domestic scene. Another example is the emerging pan-American free-trade area, commencing with the US–Canada–Mexico agreement. An integrated approach to telecommunications is starting with the Acapulco Declaration, agreed in 1992, which empowers the Organisation of American States to promote regional telecommunications development and integration, harmonisation on the region's standards and a greater uniformity of regulatory policies and pricing principles. The evolving Asia–Pacific region is also one where economic relationships are intensifying, and telecommunications co-operation amongst the ASEAN countries (Thailand, Indonesia, Singapore, Malaysia and the Philippines) is likely.

Additionally, any service provider which wishes to compete globally, whether to increase its revenue or to protect its revenue from its multinational customers by meeting their needs on a world-wide basis, will need to understand the regulatory framework in those countries in which it wishes to operate. Also, the impact of competition and domestic regulation is eroding the asymmetry of international correspondent tariffs and this is being exploited by new third party entrants which offer cheap calls by call-back from the country with the lowest tariff. This is obviously distorting the traffic balance between countries.

15.5.2 World Trade Organisation

With the rapid developments in applications of technology and the global nature of telecommunications, there is a need for a global regulator of some sort. The ITU-T is a standardising body and has no powers of regulation. In this context it is perhaps worth considering the role of World Trade Organisation and its possible involvement in the future in the regulation of global services, e.g. the Internet.

General Agreements on Tariffs and Trade (GATT) negotiations commenced in 1947 and made a significant contribution to the rapid economic growth in the 1950s and 1960s. Since the UK became a member of the EU, it no longer negotiates in the GATT on its own behalf; in matters of foreign trade, member countries have pooled their sovereignty, and the Commission of the European Communities negotiates on behalf of all member countries under a mandate from the Council of Ministers. This requires the Community line to be negotiated first but, when a common position is reached, the EU is, with the USA and Japan, one of the most influential parties.

The GATT negotiations, called the Uruguay Round, were launched in Punda del Este in 1986 and have on a number of occasions seemed near collapse. Significantly, trade in telecommunications service was included in the Uruguay Round (the Group of Negotiations on Services (GNS) telecommunications annex) as an extension of the key GATT principle of 'most favoured nation' (MFN) status. Under this concept, a country concluding a bilateral agreement with a favoured nation, giving access to the domestic market, agrees to allow any GATT member access on the same terms. Led by the USA, the more liberalised countries blocked this initiative. This caused the FCC to defer a decision on C&W's application for a licence to operate international services from the USA to the UK, on the grounds that it would weaken the USA's negotiating position in GATT.

The April 1996 round of negotiations failed to reach consensus when a group of countries led by the USA insisted in keeping mobile satellite systems outside the scope of the agreement on liberalisation. However, in February 1997, a new basic services-liberalisation pact was agreed by which about 70 governments have, in principle, consented to liberalise facilities-based access to their basic services markets, including foreign-company equity participation of up to 100%. The USA, Europe and Japan (accounting for more than 70% of global revenues) will open their markets in January 1998. Other countries will follow in the period up to 2003. However, a number of countries are seeking exemptions to the MFN principle e.g. the USA for transmission of direct broadcast satellite (DBS) television services; Brazil for the distribution of radio or television programming direct to consumers; Argentina for fixed satellite services; Bangladesh, India Pakistan, Sri Lanka and Turkey for differential measures, e.g. accounting rates, in bilateral agreements with other operators or countries. The agreement will also result in the establishment of independent regulatory bodies, in signatory countries, to foster fair competition.

15.6 Profile of an ideal regulator's role on matters of quality

Based on earlier discussions one can derive the role profile of an ideal regulator. A regulator should fulfil the best interests of the customers, suppliers, equipment manufacturers, the government and other relevant bodies national and international. As the telecommunications industry is still in the growth stage of the 'product life cycle', the profile of an ideal regulator in a few years' time could be vastly different from that of today. However, some of the general requirements and their obligations towards QoS can be postulated confidently. Such a profile would include the following characteristics:

(i)   The regulator would be a facilitator in the telecommunications industry in relation to the following:

   • awareness of the quality implications of technological advances and new services;
   • liaison with standardising bodies for the development of standards for services at the right time, i.e. influencing the prioritising of standards particularly for the customers' benefit;
   • management of assessment of service providers' delivered quality data;
   • development of an overall strategy for matters on quality;
   • developing and applying a formula for reward of punitive action against service providers based on supply of good and poor QoS.

(ii) The regulators would adopt best practice from other regulators not only in telecommunications from other countries, but also from counterparts in other industries both at home and abroad.

A successful regulator would keep up to date on technical advances in telecommunications not only in its home country but the rest of the world. It would also keep itself up to date on the implications for services It would assess its implications for the telecommunications industry and in particular that of the home country. It would try to forecast the implications on QoS and initiate early studies to assess whether any action needed to be taken.

In the standardisation fora it would attempt to have an input based on the country's needs. It would encourage development of the right standards at the right time. It would have to liaise with other regulators if it were to have any control at all; otherwise the control would be in the hands of the service providers, which form the majority of the membership of the standardisation bodies.

The regulator would play a proactive role in the selection of parameters on which delivered quality was to be published by the providers. The regulator could also assist in their definition so that those who worked with them understood the parameters unambiguously. The principal initiators of these parameters should normally be the user groups, as discussed in Chapter 14. However in certain cases, e.g. residential and special-interest customers, regulators might have to act on their behalf as there might not always be user groups to represent their interests.

The regulator has to ensure that the performance assessments of the service providers are specified and audited. Performance assessment systems may include publication of results of delivered quality, compensation schemes, quality-sensitive price caps and differentiated price/quality offerings.

The overall approach of the regulator on matters of quality would be to attain maximum improvement in the quality of life for the people through the application of telecommunications. Regulators could consider commissioning studies into topics such as quality-sensitive price caps and identification of the allocative efficient level of quality. The regulator should take the responsibility for ensuring that, in this highly competitive industry, customers do not get the quality which represents the 'lowest common denominator' the market will withstand, but the best that value technology can offer to customers.

An ideal regulator would liaise with other telecommunications regulators in other parts of the world. It would also liaise with regulators from other industries both from home and abroad. It will pick out the best practices and translate these to relevant application in the home country. If any of these practices cannot be applied within the existing framework it would put these forward to the government (which is assumed to be the body responsible for the 'citizens' charter') for consideration at the next review of the regulatory framework.

15.7 Conclusions

There appears to be a need for the role of regulator in the telecommunications industry. Regulators in the telecommunications industry, being recent arrivals, lack the experience of an established discipline e.g. law or accountancy. This, together with the fact that telecommunications is very dynamic and rapidly changing, requires the role of the regulator to be reviewed regularly. They also need to regulate competition by, for example, developing and applying a method for reward and punishment for good or bad delivery QoS by the suppliers. Meeting this challenge lies with the professionals working in the telecommunications industry, perhaps by influencing the role of the regulator.

15.8 References

1   BOWDERY, J.: 'Quality regulation and the regulated industries'. Centre for the Study of Regulated Industries, Research centre of the Chartered Institute of Public Finance and Accountancy, discussion paper, 1994
2   ROVIZZI, L., and THOMPSON, D.: 'Price-cap regulated public utilities and quality regulation in the UK'. London Business School, Centre for Business Strategy, working paper, 111, November 1991
      ROVIZZI, L., and THOMPSON, D.: 'The regulation of product quality in the public utilities and the citizen's charter', Fiscal Studies, 1992, (13:3), pp. 84–85
3   LYNCH, G. J. JUN., BUZAS, T. E., and VAN BERG, S.V.: 'Regulatory measurement and evaluation of telephone service quality', *Management Sci.*, 1994, **40**, pp. 169–194
4   ANDERSON, N. H.: 'Methods of information integration theory' (Academic Press, New York, 1982)
5   LOUVIERE, J. J.: 'Hierarchical information integration: A new method for the design and the analysis of complex multiattribute judgement problems' *in* KINEAR, T. (Ed.): 'Advances in consumer research' (Association for Consumer Research, Provo, UT, 1984)
6   LOUVIERE, J. J., and GAETH G. J.: 'Decomposing the determinants of retail facility choice using the method of hierarchical information integration: a supermarket illustration', *J. Retailing*, 1987, (63), pp. 25–28
7   LYNCH, J. G., JUN.: 'Uniqueness issues in the decomposional modelling of multiattribute overall evaluations: an information integration perspective', *J. Marketing Res.*, 1985, **22**, pp. 1–19
8   BOWMAN, E. H.: 'Consistency and optimality in managerial decision making', *Management Sci.*, 1963, **9**, (2), pp. 310–321
9   EDWARDS, E.: 'How to use multiattribute utility measurement for social decision making', *IEEE Trans.*, 1977, **SMC-7**, pp. 326-340
10   'Telecommunications service quality,' National Regulatory Research Institute, the Ohio State University, Columbus, Ohio, USA, report NRRI 96-11, 1996
11   NAOM, E. M.: 'The quality of regulation in regulating quality: A proposal for an integrated incentive approach to telephone service performance' *in* EINHORN, M. (Ed.): 'Price caps and incentive regulation in telecommunications' (Kluwer Academic Publishers, 1991), pp. 168–189
12   Telecommunications Act 1984 HMSO, London
13   MILNE, C.: 'Regulating quality of service', *in* MELODY, W. H. (Ed.): 'Telecom reform: principal policies and regulatory practices' (CTI, Copenhagen, 1997)
14   OFTEL consultative document (UK Office of Telecommunications, 1986)
15   MORI poll conducted for the NCC (National Consumer Council, London, 1987)
16   Competition and Service (Utilities) Act 1992
17   'Making service the competitive battlefield', Global Telcoms Business, **10**, June/July 1995
18   'Customer care special' supplement to *Telephony*, 6 Nov., 1995
19   WEIKLE, J. L.: 'Open your eyes to wise guys', *Rural Telecom.* Sept./Oct., 1995, pp. 13 16

20 LAWTON, R. W., ROSENBERG, E. A., MARVEL, M., and ZEARFOSS, N.: 'Measuring the impact of alternative regulatory pricing reforms in telecommunications' (NRRI, Columbus, 1994)

Additional reading list

21 BARNES, F.: 'Quality regulation: the UK experience of regulating BT', *Consumer Policy Rev.*, 1992, **2**
22 SPENCE, M. A.: 'Monopoly, quality and regulation', *Bell J. Economics*, 1975, **6**, pp. 417–429
23 VICKERS, J., and YARROW, G.: 'Privatisation: an economic analysis' (MIT, 1988)
24 HOGBEN, D.: 'Telecommunications regulation in the UK', Structured Information Program 16.1, *Br. Telecom. Eng.* (Journal of the Institution of British Telecommunications Engineers)

Chapter 16
Comparisons of quality of service

16.1 Introduction

Comparison of QoS is sometimes necessary within a country or internationally. Comparisons are important for the user, regulator, service provider and network provider. The user needs to compare delivered performances of various providers within the country and to a limited extent among international providers. The business user sometimes needs to compare the offered quality both nationally and internationally. The service provider finds it necessary to compare its offered and delivered performance with that of the competitors'. It could also benefit from comparing customer perceptions of its services with those of competitors'. This chapter gives a brief review of the existing comparisons, identifies the limitations and proposes solutions to make meaningful comparisons possible.

16.2 Categories of comparisons

16.2.1 General

Table 16.1 show the various possibilities for comparisons of performance. The principal benefits to the main beneficiaries are examined. These are dealt in Section 16.2.2 for the national comparisons and in Section 16.2.3 for the international comparisons.

Table 16.1 Possibilities of comparisons among various QoS viewpoints nationally and internationally

	Customers' QoS requirements	Service providers' offered QoS	Service providers' delivered QoS	Customers' perception of QoS
Nationally	No	Yes	Yes	Yes
Internationally	Yes	Yes	Yes	Yes

16.2.2 Comparisons within a country

Table 16.2 identifies the parties which may benefit from comparisons of the four viewpoints of quality of service.

Table 16.2 Comparisons of QoS — national scene: parties which may benefit

	Customers	Regulators	Service providers	Network providers
Customers' QoS requirements	No	No	No	No
Service providers' offered QoS	Yes	Yes	Yes	Yes
Service providers' delivered QoS	Yes	Yes	Yes	Yes
Customers' perception of QoS	No	Yes	Yes	Yes

Offered QoS
Comparison of offered QoS would benefit customers, regulators, service providers and network providers. Customers can assess whether their telecommunications needs can be met by the service quality offerings of the various providers. The regulator can assess whether any further action needs to be taken for the improvement of quality, even though such decisions have tended, in the past, to be based on the delivered quality. The service and network providers could benefit from the comparisons from a competitive angle. Such comparisons would enable providers to alter their offerings, if necessary.

It is necessary for the service providers (which could be independent of the network providers) to be aware of the network-access performance at the interface with the network providers. The national regulator would, in certain cases, stipulate guidelines on access and the interface specifications. Quality would be one of the principal considerations in such specifications.

Delivered QoS
Users and regulators can compare how well individual providers have performed relative to each other. If the comparison is sufficiently detailed, a variety of users and user groups, such as business users, special interest groups and residential users, could study the performances from their viewpoints and determine which provider was best in terms of quality. Caution must be exercised in comparing achieved performance. For example the 'average' can mask a bad 'tail' or a 'black spot' (see Section 17.7.1). Where there is reason to suspect that the average does not reflect the true state of affairs, data must be provided for the spread of results. Suitable explanations should

identify the black spots or areas of poor performance. The achieved QoS must be provided either by the service provider, with audit by an independent body, or by an independent body which has scrutinised the collection of performance data (see Section 7.4.4).

For the provider, delivered-quality data would enable comparisons to be made with its competitors. Such comparisons would highlight the need to review quality-improvement strategies.

Perceived QoS
Principal beneficiaries of customer-perception ratings are regulators and providers. These bodies can see whether customer perceptions match the delivered performance of providers and, additionally, whether there is a difference between customer perceptions of different providers. If two providers deliver similar quality, customer ratings should be similar. However, since perception is subjective (see Section 8.2.2), there could be variations in customers' perceptions of providers. Should such variations exist, these comparisons will permit their identification and allow any necessary investigation or study to be initiated.

16.2.3 International comparisons

Table 16.3 illustrates the parties which may benefit from comparisons of the four viewpoints of QoS in different countries. Matters discussed for national comparisons (Section 16.2.2) would broadly apply for international comparisons. Additional aspects applicable to the international comparisons are mentioned in this section.

Table 16.3 Comparisons of QoS — international scene parties which may benefit

	Customers	Regulators	Service providers	Network providers
Customer's QoS requirements	No	No	Yes	Yes
Service providers' offered QoS	Yes	Yes	Yes	Yes
Service providers' delivered QoS	Yes	Yes	Yes	Yes
Customers' perception of QoS	No	Yes	Yes	Yes

Customers' QoS requirements
Comparisons of performance requirements of customers could be of benefit to service and network providers, as these parties will be associated with their

counterparts in another country for the provision of international services. Comparisons of quality requirements of the users in the respective countries could be an invaluable input to the process of establishing a common and agreed level of performance, between the two countries.

Offered QoS
The principal additional beneficiaries from national comparisons are multinational customers which have private or leased networks in different parts of the world. These companies would benefit by being able to compare what is available in the market from different providers.

Network providers and service providers could also benefit from comparisons. International performance comparisons of networks could enable service providers to choose, on performance criteria, with which carrier companies they may wish to associate for the provision of international services.

16.2.4 Benchmarking

Benchmarking has been dealt with in some detail in Section 6.5.1. Comparison of performance of competitors within a country and with counterparts in another country, with a view of establishing best performance, is another benefit arising from comparisons of quality.

The important criteria for benchmarking are performance parameters defined and accepted internationally. In their absence one is left to infer meaning from incompatible data. The case of common set of definitions is made in Chapter 19.

16.3 National comparisons

16.3.1 General
The onset of liberalisation and competition in many countries has resulted in a national requirement for the publication of delivered-performance data. The publicly available comparative QoS information is reviewed here for three countries, the UK, Australia and the USA.

16.3.2 The UK

In the United Kingdom the regulatory body, the Office of Telecommunications (OFTEL), has worked with the service providers to arrive at a set of QoS performance parameters on which achieved performance is to be reported [1]. These parameters are:

(i) *Service provisioning:* percentage of orders completed on or before the date confirmed or contracted with the customer. This is broken down into:

(a)   residential (switched) service;
(b)   business switched service; and
(c)   business dedicated services (i.e. private services).

(ii) *Customer-reported faults:* customer-reported faults per 100 direct customer lines per quarter. These are broken down to the same categories as for (i) above.

(iii) *Service restoration:*

(a)   percentage of fault reports cleared in objective time; or
(b)   percentage of nonappointed fault reports cleared in objective time; and
(c)   percentage of fault reports appointed (where this can be determined).

(iv) *Complaints handling:* percentage of complaints resolved within 20 working days (a single measure covering all complaints).

(v) *Bill accuracy:* number of bill accuracy complaints received per 1000 bills issued (a single measure covering all services).

Fifteen or more service providers are currently (mid 1997) gathering achieved-performance data for the above parameters. An auditing team will vouch for the veracity of the reported results and for their consistency. The first set of such results was published in early 1996 for the last quarter of 1995.

It is expected that the list of parameters will be increased to include technical ones. At present the publication is on a voluntary basis. It is, however, expected that *all* service providers, will, in time adhere to the voluntary code and publish achieved results.

16.3.3 Australia

In Australia formal reporting of service providers' delivered performance commenced in 1994. The choice of parameters has changed with time. Continuous revision to reflect customer requirements has necessitated changes in the parameters. For example, in 1993 there were more than 50 parameters on which Telstra had to report its delivered performance. In December 1996 this has been reduced to 25. The choice of parameters to be reported by Optus, the other main service provider, reflects the services offered by it. The regulator tailors the set of performance parameters to reflect the range of services the company can offer. This contrasts with the UK practice of going for the greatest number of parameters which all providers can deal with.

The parameters on which Telstra and Optus are required to report delivered performance as of December 1996, are as follows:

Telstra
Provision of service
1   Percentage of customers connected to new services on or before the ACD[†]

---

† ACD = agreed commitment date

2   Percentage of customers connected to 'in-place services't

3   Customer access lines installed (in millions of lines)

Restoration of service, and service difficulties

4   Percentage of faults cleared within one working day of notification

5   Percentage of faults cleared within two working days of notification

6   Percentage of calls entering the service-difficulties queue answered within 15 seconds

7   Percentage of calls entering the service-difficulties queue leaving without being answered

Public switched telephone network

8   Local-call connection: percentage network loss

9   Long-distance-day-call connection: percentage network loss

10   Long-distance-night-call connection: percentage network loss

11   International-call connection: percentage loss in Telstra's network

Customer complaints

12   Total number of first-level customer complaints — reported to Telstra Business Structure

13   Total number of second-level customer complaints — received by the Telstra Business Structure

Operator-assisted services

14   Percentage of directory-assistance calls entering the network answered within 10 seconds

15   Percentage of directory-assistance calls entering the network leaving without being answered

16   Percentage of operator-assisted national calls entering the network being answered within 10 seconds

17   Percentage of operator-assisted national calls entering the network leaving without being answered

18   Percentage of operator-assisted international calls entering network being answered within 10 seconds

19   Percentage of operator-assisted international calls entering the network leaving without being answered

Payphone services

20   Public payphones — average hours to clear a fault

21   Percentage of public payphones operating at any one time

22   Percentage of public-payphone faults cleared within one working day

23   Percentage of public-payphone faults cleared within two working days

24   Public payphones — trouble reports per unit per month

---

† the path from the exchange to the customer's service delivery point still exists after a previous service cancellation

Access to itemised billing
25   Percentage of telephone lines provided with itemised billing

Optus
Fault reports and fault clearance
1   Total number of customer contacts regarding fault reports made to Optus' Customer Service Centre
2   Percentage of customer contacts regarding fault reports closed on initial contact
3   Percentage of faults cleared within one day

National and international long-distance services
4   Network loss on national long-distance calls within the Optus network as a percentage of call attempts (not including congestion)
5   ABR performance for five frequently called international destinations

Customer satisfaction
6   Customer-satisfaction index for residential long-distance services (score out of 100)
7   Customer-satisfaction index for residential mobile services (score out of 100)
8   Customer-satisfaction index for business long-distance services (score out of 100)
9   Customer-satisfaction index for business mobile services score out of 100)

Service availability
10   Percentage of the population offered access to Optus' long distance services

Complaints
11   Total number of complaint received by Optus' Customer Service Centre

Mobile-telephone-network service parameters
The delivered performance of mobile telephone services is to be published based on the following parameters, for each of the eight states of Australia:

Analogue mobile phone services
AMPS call drop out (%)
AMPS call congestion (%)

Global system for mobile communications
GSM call drop-out (%)
GSM call congestion (%)

Two new Network Standards were published by the Australian Telecommunications Authority (AUSTEL) in 1996:

Technical Standard 027, which sets performance criteria for public networks in terms of call success rates and transmission quality; and

Technical Standard 029, which sets levels of required overall accuracy of network licensees' charging and billing systems.

The two fixed-network carriers, Telstra and Optus, are required to demonstrate that they have complied with these standards. The mobile-service carrier Vodafone and the fixed-service provider AAPT have to comply with the second standard. All four licensees are expected to undertake test calling and other compliance activities during the April-June 1997 quarter and subsequently provide results. These are additional requirements.

16.3.4 United States of America

In the USA there are two sets of 'national' QoS statistics to be published: the first is the requirement from the Federal Communications Commission (FCC) and the second is the local state requirement. The move towards publication of the QoS of local operating companies in the USA started with the divestiture of AT&T of its local services. Over the years, improvements were made to meet improved operational requirements and currently a quarterly return is required from each principal local operating company. The parameters on which the FCC requires regular publication of QoS statistics are illustrated here for the last quarter of 1995. The list of parameters required by Florida is shown to illustrate the requirements of a state.

The regional Bell operating companies (RBOCs) report on the following parameters twice a year, as required by the FCC. The reports are compiled aggregated to the holding-company level. A holding company usually serves more than one state.

(i) percentage of customers satisfied — residential
(ii) percentage of customers satisfied — small businesses
(iii) percentage of customers satisfied — large business
(iv) percentage of offices providing dial tone in less than 3 seconds.

The first three are indicators of customers' perception.

The RBOCs and the other companies report quarterly on the following achieved performance (at the holding company level):

Access services provided to carriers — switched access
1 Percentage installation commitments met
2 Average missed installation (days)
3 Average repair interval (hours)

Access services provided to carriers — special access
4 Percentage installation commitments met
5 Average missed installation (days)
6 Average repair interval (hours)

Local services provided to residential and business customers
7 Percentage of installation commitments met

8   Percentage of installation commitments met — residence
9   Percentage of installation commitments met — business
10   Average missed installation (days)
11   Average missed installation (days) — residence
12   Average missed installation (days) — business
13   Initial trouble reports per thousand lines
14   Initial trouble reports — total Metropolitan Statistical Area (MSA)
15   Initial trouble reports — total non-MSA
16   Initial trouble reports — total residence
17   Initial trouble reports — total business
18   Troubles found per thousand lines
19   Repeat troubles as a percent of trouble reports
20   Repeat troubles — total residence
21   Repeat troubles — total business

Customer complaints per million access lines
22   Customer complaints — residential
23   Customer complaints — business
24   Total access lines in thousands
25   Total trunk groups
26   Total switches

Switches with downtime
27   Number of switches
28   As a percentage of total switches

Average switch downtime in seconds per switch
29   For all occurrences or events
30   For unscheduled events over 2 min

For unscheduled downtime more than 2 min
31   Number of occurrences or events
32   Events per million access lines
33   Average outage duration, in minutes
34   Average lines affected per event in thousands
35   Outage line-minutes per event in thousands
36   Outage line-minutes per 1000 access lines

For scheduled downtime more than 2 min
37   Number of occurrences or events
38   Events per million access lines
39   Average outage duration, minutes
40   Average lines affected per event in thousands
41   Outage line-minutes per event in thousands
42   Outage line-minutes per 1000 access lines
43   Percent of trunk groups exceeding blocking objective 3 months

Total number of outages
44 Scheduled
45 Procedural errors — telephone company (installation; maintenance)
46 Procedural errors — telephone company (other)
47 Procedural errors — system vendors
48 Procedural errors — other vendors
49 Software design
50 Hardware design
51 Hardware failure
52 Natural causes
53 Traffic overload
54 Environmental
55 External power failure
56 Massive line outage
57 Remote
58 Other/unknown

Total outage line-mounted minutes per thousand access lines
59 Scheduled
60 Procedural errors — telephone company (installation; maintenance)
61 Procedural errors — telephone company (other)
61 Procedural errors — system vendors
62 Procedural errors — other vendors
64 Software design
65 Hardware design
66 Hardware failure
67 Natural causes
68 Traffic overload
69 Environmental
70 External power failure
71 Massive line outage
72 Remote
73 Other/unknown

The delivered performance of the following parameters are required by the Florida Public Service Commission.

(a) Dial-tone delay
1   Dial-tone delay

(b) Call completion
2   Intraoffice
3   Interoffice
4   Extended Area Service (EAS, a service between 'local and long distance' offered in certain geographical areas)
5   Interlata DDD (lata=local access transport area, DDD=direct distance dialling)

(c)   Incorrectly dialled calls
6   Incorrectly dialled calls

(d)   911 service
7   911 calls

(e)   Transmission
8   Dial-tone level
9   Central-office loss
10   MW frequency (a local-transmission-related terminology)
11   Central-office noise — metal
12   Central-office noise — impulse
13   Subscriber loops

(f)   Power generators
14   Power generators

(g)   Test numbers
15   Test numbers

(h)   Central office
16   Scheduled routine programme
17   Frame
18   Facilities

(i)   Answer time
19   Operator
20   Directory assistance
21   Repair service
22   Business office

(j)   Adequacy of directory and directory assistance
23   Directory service
24   New numbers
25   Numbers in directory

(k)   Adequacy of intercept services
26   Changed numbers
27   Disconnected service
28   Vacation disconnects
29   Vacant numbers
30   Disconnect nonpay

(l)   Toll timing and billing accuracy
31   Intra-lata timing accuracy
32   Intra-lata over time
33   Intra-lata under time
34   Credit card-intra time
35   Credit card-intra under time
36   Directory-assisted bill accuracy

(m) Public telephone service
37 Payphone/exchange
38 Serviceability
39 Handicapped access
40 Glass
41 Doors
42 Level
43 Wiring
44 Cleanliness
45 Lights
46 Telephone numbers
47 Name or logo
48 Dial instructions
49 Transmission
50 Dialling
51 Coin return auto
52 Coin return operator
53 Operator ID (identification) of coins
54 Access all LD carrier (long distance)
55 Ring back operator
56 Coin-free access to operator
57 Coin-free access directory access
58 Coin-free access to 911
59 Coin-free access repair service
60 Coin-free access to business office
61 Directory
62 Directory security
63 Address/location

(n) Availability of service
64 3-day primary service
65 Primary service appointment

(o) Repair service
66 Restored same day
67 Restored 24 hour
68 Repair appointments
69 Rebates over 24 hour
70 Service affecting 72 hours

(p) Customer complaints
71 Justified complaints/1000 lines

The rather detailed reporting of availability of access and of outages probably reflects the needs of the users. Every country tends to publish performance parameters to reflect its local needs and the definition of parameters would therefore be different in various countries and not reflect any international comparisons requirement.

16.4 International comparisons

16.4.1 Member countries of OECD

The aim of the performance-indicator group within the OECD[†] is to publish comparable performance indicators from the member countries. Attempts are being made to arrive at a commonly agreed set of definitions of performance to be used by all member countries to report achieved performance. The OECD is unlikely to develop an architectural framework on QoS, as it does not appear to consider this task to be in its remit. However, its collective performance reports from the member countries may be considered useful.

Data on comparisons of performance in the international arena (apart from OECD figures) have been available, albeit of limited value, for at least two decades. Comparisons were originally initiated by the service providers at a time when liberalisation and competition were either nonexistent or limited to the USA. The comparison which is currently of most value and publicly available is published by the OECD [2]. Other comparisons are either for use within PTOs or are of limited circulation. The OECD has been carrying out surveys among its member countries on the importance of QoS for telecommunications. Respondents have indicated that relative priorities depends on the state of their own telecommunications environment. For example, in the Scandinavian countries, where telephone penetration was already very high, priority for unmet demand was very low. In countries where unmet demand is high the priority for reducing the waiting list will, naturally, be high. The QoS parameters on which the OECD has published achieved performance, in 1997, but for a previous year, 1995, are as follows:

1   Waiting time for connection of new service
2   Number of outstanding connections
3   Number of payphones
4   Percentage of payphones which are cardphones
5   Number of payphones per 1000 people
6   Average percentage of payphones in working order
7   Call-failure rates for local and long-distance connections
8   Faults per 100 lines per annum
9   Faults repaired within 24 hours
10   Customers served with itemised billing

†The Organisation for Economic Co-operation and Development (OECD) was formed in 1961 to succeed the Organisation for European Economic Co-operation (OECC) which had been established in 1948 to implement the Marshall Plan. Its aims are to promote economic and social welfare throughout the OECD area. The 29 member countries are Australia, Austria, Belgium, Canada, Czech Republic, Denmark, Finland, France, Germany, Greece, Hungary, Iceland, Ireland, Italy, Japan, Republic of Korea, Luxembourg, Mexico, Netherlands, New Zealand, Norway, Poland, Portugal, Spain, Sweden, Switzerland, Turkey, United Kingdom and United States of America.

11 Customers served with caller-line identification
12 Charges for directory assistance
13 IDD completion rates in the area

The fault rate per 100 lines per annum and the fault repair rates are reproduced in Table 16.4 to illustrate what was achieved.

The principal shortcomings of the OECD statistics on quality are:

(i) for some parameters comparative data are not to commonly agreed definitions;
(ii) comparative data are not on a service-by-service basis except in a few instances;
(iii) not all member countries report all the performance figures;
(iv) the OECD publication is made once in two years; therefore many of the results are dated. The 1997 publication contained figures applicable for the 1995 year.

Table 16.4 Performance of OECD member countries on fault incidence and repair time, as of 1995 returns

Country	Fault per 100 lines per year	Fault repair within 24 hours	Notes
Australia	na	67.0	
Austria	19.0	93.0	Including CPE faults
Belgium	2.2	87.0	
Canada	2.3	85.4	Bell Canada only; repair times are for residential
Czech Republic	10.7	90.3	Within 72 hours
Denmark	na	86	Within 72 hours
Finland	8.3	69.1	In a working day
France	6.3	88.3	
Germany	8.7	93.0	Within 3 working days
Greece	43.4	58.4	
Hungary	39.5	78.1	
Iceland	n.a.	n.a.	
Ireland	17.0	n.a.	
Italy	12.6	92.3	Excluding CPE
Japan	1.7	100.0	Fault-repair percentage is approximate
Luxembourg	n.a.	n.a.	
Mexico	4.6	78.8	
Netherlands	2.5	87.0	Within 48 hours
New Zealand	41.0	73.0	Telecom New Zealand Residential only

Norway	14.0	73.7	Within 8 hours
Portugal	38.0	91.0	Within two working days
Spain	n.a.	n.a.	
Sweden	8.4	85.0	Telia AB residential only
Switzerland	14.0	94.0	
Turkey	60.2	95.0	
United Kingdom	14.0	82.0	BT only, within 5 or 9 working hours
United States	16.9	n.a.	

Data from Poland and Republic of Korea were not included as these countries became members of OECD in late 1996.

16.4.2 European Union member countries

The European Commission intends the member countries of the European Union (EU) to publish delivered performance for the principal services provided by the telecommunications-service providers from the member countries. To date (1997) four documents have been published where some quality requirements have been stated. These are:

- Recommendation on ISDN
- Directive on leased lines
- Recommendation on packet switched data service (PSDS)
- Directive on voice telephony

A Recommendation is not legally binding to the member countries of the EU, whereas a Directive is.

ISDN: Recommendation 92/383/EEC dated 5 June 1992
The following QoS parameters are recommended for the ISDN in the above Recommendation:

(i)   availability of access;
(ii) mean time between interruptions;
(iii) bit-error ratio;
(iv) connection-processing delay;
(v)   network transit delay;
(vi) unsuccessful-calls ratio;
(vii) indicators for packet-mode bearer services as for PSDS (given in this subsection).

Since the document has the status of a Recommendation, publication of achieved performance of the above parameters are not mandatory.

Leased lines: Council Directive 92/44/EEC dated 5 June 1992
Article 4 of the above document list the following information to be provided by the service providers:

(i) typical delivery period, which is the period, counted from the date when the user has made a firm request for a leased line, in which 80% of all leased lines of the same type have been put through to the customers;
(ii) the contractual period, which includes the period which is in general foreseen for the contract and the minimum contractual period which the user is obliged to accept;
(iiii) the typical repair time, which is the period counted from the time when a failure message has been given to the responsible unit within the service provider up to the moment in which 80% of all leased lines of the same type have been re-established and in appropriate cases notified back in operation to the users;
(iv) any refund procedure.

In addition, the quality-related matters listed in Table 16.5 are also mentioned in Annex II of the above document. These are to take effect from around late 1997.

Table 16.5   Quality-related matters referred to in Annex II of Directive 92/44/EEC

Leased-line type	Interface-presentation specifications	Connection-characteristics performance specifications
Ordinary-quality voice bandwidth, analogue	2 wire[1]: ETS 300 448[3] or 4 wire[2]: ETS 300 451[4]	2 wire: ETS 300 448[3] 4 wire[2]: ETS 300 451[4]
Special-quality voice bandwidth, analogue	2 wire[1]: ETS 300 449[5] or 4 wire[2]: ETS 300 452[6]	2 wire[1]: ETS 300 449[5] 4 wire[2]: ETS 300 452[6]
64 kbis/s digital[7], unstructured	ETS 300 288 ETS 300 288/A1[8]	ETS 300 289
2048 kbit/s digital, unstructured[9]	ETS 300 418	ETS 300 247 ETS 300 247/A1
2048 kbit/s digital, structured[10]	ETS 300 418[11]	ETS 300 419[12]

Notes

1   The attachment requirements for terminal equipment to be connected to these leased lines are described in Common Technical Regulation 15 (CTR 15).
2   The attachment requirements for terminal equipment to be connected to these leased lines are described in Common Technical Regulation 17 (CTR 17).
3   Until 31 December 1997 these leased lines may be provided in accordance with ITU-T Recommendation M.1040 (1988 version) instead of ETS 300 448.
4   Until 31 December 1997 these leased lines may be provided in accordance with ITU-T Recommendation M.1040 (1988 version) instead of ETS 300 451.
5   Until 31 December 1997 these leased lines may be provided in accordance with ITU-T Recommendation M.1020/M.1025 (1988 version) instead of ETS 300 449.
6   Until 31 December 1997 these leased lines may be provided in accordance with ITU-T Recommendation M.1020/M.1025 (1988 version) instead of ETS 300 452.

7   The attachment requirements for terminal equipment to be connected to these leased lines are described in Common Technical Regulation 14 (CTR 14).
8   For an interim period extending beyond 31 December 1996, these leased lines may be provided using other interfaces, based on X.21 or X.21 bis, instead of ETS 300 288.
9   The attached requirements for terminal equipment to be connected to these leased lines are described in Common Technical Regulation 12 (CTR 12).
10 The attached requirements for terminal equipment to be connected to these leased lines are described in Common Technical Regulation 13 (CTR 13).
11 Until 31 December 1997 these leased lines may be provided in accordance with ITU-T Recommendation G.703, G.704 (excluding section 5) and G.706 (cyclic redundancy checking) (1988 version) instead of ETS 300 418.
12 Until 31 December 1997 these leased lines may be provided in accordance with relevant G.800 series ITU-T Recommendations (1988 version) instead of ETS 300 419.

For the types of leased lines listed above, the specifications referred to also define the network termination points (NTPs), in accordance with the definitions given in article 2 of Directive 90/387/EEC.

Packet Switched Data Service – Recommendation 92/382/EEC dated 5 June 1992
The following parameters have been recommended:

(i)   Unsuccessful calls due to network congestion (NC) calls;
(ii) Service availability;
(iii) Mean time between NC disconnection;
(iv) Transmitted throughput;
(v)   Received throughput;
(vi) Round-trip delay;
(vii) Call-set-up delay.

The document has the status of a Recommendation; publication of performance of parameters are not therefore mandatory.

Voice telephony – Directive 95/62/EC dated 13 December 1995
Table 16.6 lists the QoS parameters on which member countries ought to publish delivered performance. These telephony-quality-related criteria are being revised, and are expected to be formally adopted towards the end of 1997.

Table 16.6 QoS parameters specified in Directive 95/62/EC

Indicator[1]	Definition	Measurement method
1 Supply time for initial connection	ETSI ETR 138	ETSI ETR 138
2 Fault rate per access line	ETSI ETR 138	ETSI ETR 138
3 Fault-repair time	ETSI ETR 138	ETSI ETR 138
4 Unsuccessful-call ratio	ETSI ETR 138	ETSI ETR 138
5 Call-set-up time	ETSI ETR 138	ETSI ETR 138
6 Response times for operator services	ETSI ETR 138	ETSI ETR 138
7 Response times for directory enquiry services	as for operator services	as for operator services
8 Proportion of coin – and card-operated public paytelephones in working order	ETSI ETR 138	ETSI ETR 138
9 Billing accuracy	(see note 2)	(see note 2)

Notes

1   Indicators should allow for performance to be analysed at a regional level (i.e. no less than level 2 in the Nomenclature of Territorial Units for Statistics (NUTS) established by Eurostate).
2   National definitions and measurement methods should be used until such time as a common definition and measurement methods are agreed at the European level.

16.5 Management of comparisons of QoS

16.5.1 Structured approach

A methodology for identifying and addressing the key issues is illustrated in Figure 16.1

The key issues to be addressed are:

(a) whether the comparisons are to be carried out on a national or international level, and for what viewpoints;
(b) whether data for comparisons are to be on a service-specific basis or based on a basket of services;
(c) development of definitions of parameters;
(d) establishment of the logistics of producing performance data;
(e) establishment of an audit process; and
(f) publication arrangements.

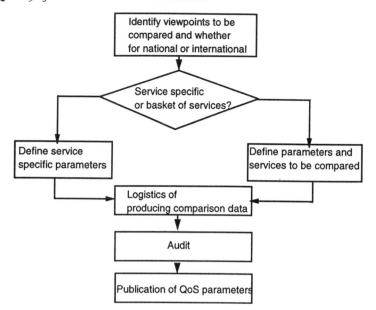

Figure 16.1   Schematic diagram for the management of comparisons of quality of service

16.5.2 Determination of viewpoints

The first task is to identify whether comparisons are to be carried out at a national or international level. At a national level, comparisons of customers' QoS requirements are of little value. All other comparisons shown in Figure 16.2 may be considered.

16.5.3 Service specific or basket of services?

For simple comparisons of overall performance, a basket of services may be chosen. For example, the basic telephone service with payphone availability may be considered adequate for the performance comparison. More complex baskets may be produced for the higher end of the market. At the most sophisticated level, the comparisons should be on a service-by-service basis.

16.5.4 Development of definitions

Performance parameters must be defined if these are to be compared on a service-specific basis. Such definitions should be carried out at a national level for use within a country and at an international level for international comparisons. It is not necessary for the national definitions to be agreed at an international level. However, it may be useful for parameters to be defined at an international level as this would enable service providers to provide data which can be equally used within the country and for international comparisons, without additional effort.

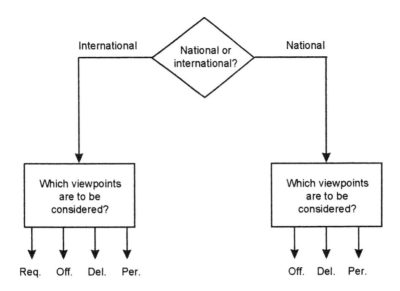

Figure 16.2   Logic diagram to determine the viewpoints for comparisons

Definitions of parameters must be unambiguous and clear. For example, parameters to illustrate time for provision of a service should clearly state whether the time is in clock hours/days or working hours/days. There are many pitfalls in arriving at unambiguous definitions, and these should to be identified by the parties which will provide and use the comparative data.

16.5.5 Logistics

The logistics of producing performance data consist of specification of monitoring systems, collection of measured performance values and the computation of end-to-end performance. These have been dealt with in Section 7.4.

16.5.6 Establishment of audit process

Performance data produced for purposes of comparisons require an audit process to add credibility. Audit processes are described in Section 7.4.4.

16.5.7 Publication arrangements

The publication of national comparisons would normally be carried out by the regulator. International comparisons would be published by a body such as the OECD. The principal issues related to publication of delivered performance have been dealt in Section 7.4.3.

16.6 Proposed parameters

As a way of illustration, a set of parameters is indicated, in Table 16.7, which may be used to indicate the achieved or delivered performance of basic telephony. These parameters may be used within a country for comparisons purposes. A subset of these may be used for international comparisons. Choice of parameters within a country would normally reflect local needs. Comparisons of international performances would depend on the collective requirements of member countries and the parameters would be chosen by the body responsible for such comparisons, e.g. the OECD or the World Trade Organisation. A standardising body such as ITU-T would define the parameters.

Table 16.7 Suggested parameters for international comparison of achieved quality of basic telephony service

1 Time for provision of service
2 Time for resolution of complaints
3 Time for repair
4 Time for connection set up
5 Measure of misrouted calls
6 Availability of service
7 Call quality
8 Cut-off during conversation phase
9 Billing accuracy

16.7 Conclusions

Current performance-comparison systems within a country appear to have evolved to suit local needs. This is particularly true in countries where competition has been in place for some time, e.g. in the USA. In other countries where competition has been in existence for a shorter period, e.g. the UK and Australia, performance parameters are still being developed by the regulators for use by the service providers. It will take some time before an optimum set of parameters has evolved on which delivered performances are required to be published.

The principal requirement for effective comparisons of performance — agreed definitions of parameters — is lacking. Within a country it is up to the local regulator to ensure that such definitions are agreed on. However, in the international scene an agreed set of performance definitions is required for effective comparisons of performance. The onus is upon the user groups and the service providers throughout the world to address this issue, perhaps through the international forum ITU-T.

In the international arena there has been little cause for any body to influence countries and their service providers either to develop a set of performance parameters commonly defined or to publish achieved results. The European Commission has made an attempt in this area; however its directives have not yet been implemented.

16.8 References

1 OFTEL 'Telecommunications companies: comparable performance indicators', published quarterly, in the UK by the Office of Telecommunications
2 OECD: 'Communications Outlook', (Organisation for Economic Co-operation and Development, 1997)

Chapter 17

Practical management of
quality of service

17.1 Introduction

In this chapter the principal tasks associated with the management of quality of basic telephony are illustrated with steps associated with the four viewpoints of the quality cycle, discussed in Chapter 4. In the illustrations the number of parameters on which customers' QoS requirements considered have been limited to 10. In practice, the actual number of parameters to be considered will depend on the degree of sophistication in the management of quality by the service provider. The principles illustrated here may be equally well applied to any sophisticated service platform, e.g. frame-relay, asynchronous transfer mode (ATM) or one-way transmission services such as television and radio broadcasting.

An illustration of customers' requirements of quality criteria for personal communications services and on the management of network congestion are given.

17.2 Determination of customers' QoS requirements

17.2.1 General

The first step in the management of QoS is the determination of customers' requirements. This involves two tasks: first, the determination of levels of performance required for the performance parameters and, secondly, analyses of these data to be put in a suitable form for submission into the next stage of the quality cycle, i.e. the determination of offered quality. Customers' QoS requirements may be determined by questionnaires, telephone interviews or personal interviews. An example of a postal questionnaire is given in Appendix 3. When telephone interview or personal interview is adopted, the questionnaire may be used as the basis for discussion with customers. The examples of performance parameters are for illustrative purposes only. Selection of parameters, in practical management of quality must reflect the customers' performance requirements and are therefore likely to vary from country to country and also depend on the technology used for the switching and transmission equipment in the network.

17.2.2 Analyses of customers' requirements

Analyses of customers' requirements consist of two components: first, prioritisation of parameters and, secondly, estimation of the performance level of every parameter to satisfy various proportions of the customer population. These two analyses are illustrated by the following examples.

Prioritisation of parameters

The average priority rating for each parameter is computed and a simple ranking is carried out. An illustration of a prioritised list is given in Table 17.1 with the average priority-rating scores.

Table 17.1 Prioritised performance parameters and average rating scores (population = 100)

Priority number	Parameter	Rating score
1	Time for provision	9.5
2	Time to repair	8.8
3	Accuracy of bills	8.6
4	Availability (network resource to establish a call)	7.5
5	Call quality	7.1
6	Misrouted calls	7.0
7	Availability (for the intended duration of call)	6.9
8	Speed of complaints handling	5.5
9	Time to establish a connection	4.5
10	Time for resolving precontract activities	3.9

Estimation of performance levels to satisfy various proportion of customer population

Time for provision: Consider the distribution shown in Table 17.2 of values quoted by 100 respondents.

Table 17.2 Distribution of respondents' replies for time for provision of telephony service

Time for provision (days)	1	2	3	4	5	6	7	8	9	10
Number of customers	3	5	7	11	20	27	18	4	3	2
Cumulative percentage	100	97	92	85	74	54	27	9	5	2

This information may be illustrated graphically as cumulative percentage versus time for provision, as shown in Figure 17.1. From this graph the following information may be deduced:

%

Figure 17.1   Customer's requirements for time to provision of a service against cumulative percentage of customers

Maximum time for provision required by
95% of the customers = 2 days
90% of the customers = 3 days
80% of the customers = 4.5 days

Time to repair: Analysing the respondents' replies for time to repair in a similar manner could give the following conclusions:

Time to repair to satisfy
95% of customers   = 6 hours
90% of customers   = 8 hours
80% of customers   = 12 hours

Accuracy of bills: This parameter has two values to express the required level of quality. One is the maximum number of errors that may be tolerated and the other is the maximum magnitude of the error. These may be treated in the same manner as the treatment for the processing of 'time for provision'. The findings may thus be illustrated as follows:

Maximum number of errors to satisfy
95% of customers   = 1
90% of customers   = 2
80% of customers   = 3

Maximum magnitude of error to satisfy
95% of customers   = 0.25 penny
90% of customers   = 0.5 penny
80% of customers   = 1 penny

Before the above figures can be put into the process for the determination of the offered QoS level, they must be expressed as a proportion or a percentage of the customers' bills. For example, if the average yearly bill is £120 for calls only (i.e. excluding rental charges), the percentage of the magnitude becomes:

Maximum magnitude of error to satisfy
95% of customers   = (0.25 penny/£120) × 100
                      = 2.1 parts in 100 000 or 0.0021%

90% of customers	$= (0.5 \text{ penny}/£120) \times 100$
	$= 4.2$ parts in 100 000 or 0.0042%
80% of customers	$= (1 \text{ penny}/£120) \times 100$
	$= 8.3$ parts in 100 000 or 0.0083%.

Assuming that 95% of customers make an average of 500 calls each per year, the requirement becomes:

Maximum tolerable number of errors = 1 in 500 calls **and** the magnitude of the error not to exceed 2 parts in 100,000 for any one error.

Availability: Two sets of values to express availability may be processed and the customers' requirements may be expressed as follows:

Maximum number of outages to satisfy
95% of customers	= 1 in a year
90% of customers	= 1.25 in a year
80% of customers	= 1.7 in a year

Maximum duration of any one outage (end-to-end, unplanned) to satisfy
95% of the customers	= 20 seconds
90% of the customers	= 25 seconds
80% of the customers	= 30 seconds

Call quality: Call-quality requirements, from the survey data, may be represented as follows:

Maximum number of calls that can be tolerated by customers for moderate difficulty in understanding person at the other end of a telephony call by
95% of customers	= 2 calls in one year
90% of customers	= 3 calls in one year
80% of customers	= 4 calls in one year

Misrouted calls: Maximum tolerable misrouting requirements, from the survey data, may be represented as follows:

Maximum number of misrouted calls that will be tolerated by
95% of the customers	= 1 call in 1000
90% of the customers	= 1.5 calls in 1000
80% of the customers	= 2 calls in 1000

Availability of connection for the intended duration of the call: Availability requirements, from the survey data, may be represented as follows:

Maximum number of calls cut off during conversation that will be tolerated by
95% of the population	= 1 call in 1000
90% of the population	= 2 calls in 1000
80% of the population	= 2.5 calls in 1000

Time for resolution of complaints: Requirements for resolution of complaints, from the survey data, may be represented as follows:

Maximum time for resolution of complaints for
95% of the customers   = 1 hour
90% of the customers   = 1.5 hours
80% of the customers   = 2 hours

Time to establish connection (after last digit has been keyed): Requirements for maximum time for connection to called party, from the survey data, may be represented as follows:

Maximum time to establish connection after keying in last digit for
95% of customers   = 1 second
90% of customers   = 1.5 seconds
80% of customers   = 2 seconds

Time for completing all precontract activities: Requirements for maximum time for completing all pre-contract activities, from the survey data, may be represented as follows:

Maximum time to deal with pre-contract activities for
95% of the customers   = 2 hours
90% of the customers   = 2.2 hours
80% of the customers   = 2.5 hours

17.3 Determination and specification of offered QoS by the service provider

17.3.1 General

Determination of offered or planned quality of service from customers' requirements and service providers' business considerations may be illustrated by the processes shown in Figure 17.2.

The 10 performance criteria in Table 17.1 are taken through the above process in Sections 17.3.2–17.3.11. The offered quality is expressed as a target and planning guidelines are drawn from these. Detailed specifications for implementation are derived from these guidelines. Discussion of the detailed specifications is outside the scope of this book.

17.3.2 Time for provision

Customers' requirements: Provision to be within two working days to meet 95% of the customers' requirements.

Current performance level: Six working days to provide service to 95% of customers seeking connection or provision of service.

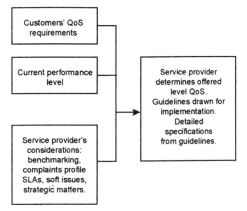

Figure 17.2   *Process for the determination of offered QoS from customers' requirements*
by service provider

Considerations:

- benchmark: corresponding best time for provision is 5 min;
- complaints profile: around five complaints per year per 100 installations, mostly on change of installation dates;
- service-level agreements: emergency services have indicated the need for tightening of service level agreements (SLAs) due to two calls which did not get though as quickly as they should have; these have resulted in increased damage to premises caused by the delay in fire services putting out the fire;
- soft issues: comments on the large number of people on site to install a telephone, sometimes up to five people including trainees; workmen dirty and patronising; lack of professional attitude to work or the customers has also been reported;
- strategic considerations: competitors are offering faster provision time. They provide the fastest provision for hospitals and emergency services with slower service for ordinary residential customers;
- cost of increasing quality: equivalent of 50 man days for the year.

Offered QoS from above considerations and planning guidelines:

- target provision time: two working days, to be achieved by the end of the following year;
- high-profile customers to be given priority for installation;
- when a date is offered to a customer for installation, it is not to be changed, unless at the request of the customer;
- workmen to be reduced in number at installation sites; all installation staff to be provided with new uniforms and sent on training course on treatment of customers in a professional manner; and
- alternative routing of emergency calls to be increased to ensure virtual 100% availability at all times.

17.3.3 Time to repair

Customers' requirements: 6 hours maximum to satisfy 95% of the customers.

Current performance level: One and half days (12 working hours) to repair 95% of the faults reported.

Considerations:

- benchmark: 2.5 hours to repair 95% of faults reported;
- complaints profile: steady trickle of complaints on too much time taken for repair;
- service-level agreements: hospitals and utilities, (fire, emergency services,) require much shorter repair time for faults;
- soft issues: repair staff said to be indifferent and appear to give the impression that they are doing a favour by repairing the fault; no call back to ask if the repair has satisfactorily carried out;
- cost of quality improvements: 10% increase in repair staff for next three years. One man year of professional staff to improve quality procedures in the system;
- strategic considerations: competitors promise 2 hours repair time for prestige addresses and utilities, but not for ordinary residential customers.

Offered QoS from above considerations and planning guidelines:

- repair time target to be maximum of 8 hours for 95% of faults reported;
- priority to be given to fault repairs for utilities, hospitals etc;
- repair staff to be sent on a training course to teach them the professional way to treat customers.

17.3.4 Accuracy of bills

Customers' requirements: To satisfy 95% of customers, not more than one error in 500 calls and the magnitude of the error not to be greater than two parts in 10 000.

Current performance level: 15 complaints per 1000 bills.

Considerations:

- benchmark: accuracy of not more than one error in 10 000 chargings;
- service-level agreements: large companies and representatives of residential customers require independent audit trails on charging and billing processes; the audit-trail findings should indicate error not more than the customers' requirements stated above; compensation required if the targets are not met;
- soft issues: bad publicity if customers' requirements are not met;
- strategic considerations: failure to meet 100% error-free billing could result in enquires from the regulator; competitors' supply better accuracy due to more modern switching and charging equipment.

Offered QoS from above considerations and planning guidelines: The service provider decides on a programme to upgrade switching and charging/billing equipment, establishment of independent audit trail and compensation if independent audits show that the customers' requirements are not met.

17.3.5 Availability of network resource to establish a call

Customers' requirements: To satisfy 95% of customers, not more than one unplanned outage a year and its duration not to be more than 20 seconds.

Current performance level: Outage has been consistently quoted in the region of 2 hours outage (unplanned) per year with a maximum duration of 15 min for any one outage.

Considerations:

- benchmark: one outage per year, duration of outage not more than 30 seconds;
- complaints profile: steady level of complaints from hospitals and emergency services notifying cases of unavailability of service and the resulting inconvenience to staff and patients;
- service-level agreements: pressure from hospitals and emergency services to improve availability and to offer SLAs with required availability;
- soft issues: none;
- strategic considerations: competitors offer availability required by the hospitals and emergency services, but not to residential customers;
- cost of improvements: 10% of the original cost of the exchange.

Offered QoS from above considerations and planning guidelines: Target performance: not more than one outage with duration of outage less than 30 min. Improvements to be carried out on the software control of the digital local exchanges as this element is considered to be weakest link in the end-to-end availability. Software enhancements to be carried out on all local exchanges during the next year. Target to be achieved in the year after upgrades have been implemented.

17.3.6 Call quality

Customers' requirements: To satisfy 95% of customers, not more than two calls in a year (or 2 in 500 calls for an average customer) should experience moderate difficulty in understanding the person at the other end.

Current performance level: Average of three calls in 100 experiencing moderate difficulty to understand the person at the other end.

Considerations:

- benchmark: one call in 1000 experiencing difficulty in understanding called or calling party;
- complaints profile: call quality not quite good enough;

- service-level agreement requirements: better level of quality to be guaranteed;
- soft issues: none;
- strategic considerations: if transmission quality is poor and competitors' performance is better, as is likely to be case with their newer equipment, business would be lost to them.

Offered QoS from above considerations and planning guidelines: Improve call quality to a level of not more than 1 call in 100 to experience moderate difficulty in understanding the called party. As most of the poor-quality calls are due to the analogue network, which is due for replacement by a digital network by the end of the next year, the target is expected to be achieved without any additional resources.

17.3.7 Misrouted calls

Customers' requirements: To satisfy 95% of customers, not more than one misrouted call in 1000.

Current performance level: Estimated proportion of misrouted calls is 1 in 10 000 for the digital network and 1 in 1000 for the analogue network.

Considerations:

- benchmark: world's best performance is estimated to be not more than one misrouted call in 10 000;
- complaints profile: no serious level of complaint;
- service-level agreement: no serious contention in this area;
- soft issues: none;
- strategic considerations: competitor's overall performance is better, due to fully digital equipment;
- cost: publicity for 12 days in the next 12 months.

Offered QoS from above considerations and planning guidelines: Customers to be advised that the digitalisation programme is to be completed by the end of next year and that the proportion of misrouted calls will then meet the required level.

17.3.8 Availability of network resource for the intended duration of call

Customers' requirements: Maximum number of calls cut off during conversation which will be tolerated by 95% of the customer population is one call in 1000.

Current performance level: An estimate of three calls in 1000.

Considerations:

- benchmark: Not more than one call in 2000 cut off during conversation;
- complaints profile: complaints of some calls, especially international, being cut off during conversation;

- service-level agreements: no contention in this area at present;
- soft issues: none at present;
- strategic considerations: competitors, using latest technology and with limited international coverage, claim better performance.

Offered QoS from above considerations and planning guidelines: The availability for network resources for the intended duration of the call is expected to be within the target of both the customers and the service provider when the digitalisation program is complete. The inability of the service provider to improve the performance of international circuits beyond the service provider's boundary should be indicated to the customers. It is probably not necessary for the situation to be monitored, as the expected level of performance will be achieved.

17.3.9 Time for resolution of complaints

Customers' requirements: Response from the service provider on what has been done or will be done to remedy the situation within a maximum of 1 hour from making the complaint, for 95% of the population.

Current performance level: Five days for the resolution of 90% of the complaints.

Considerations:

- benchmark: definitions of complaints vary from country to country and the method of resolution varies; however the world's best appears to be around one hour to clear 90% of the complaints;
- complaints profile: general theme of complaint is that it takes too long for resolution and often there is no come back; customers have to make repeat contacts for answers;
- service-level agreements: hospitals and emergency services want 15 min for resolution of complaints;
- soft issues: poor image of the service provider among customers for shoddy service and lack of respect for the customers;
- strategic considerations: competitors are much more polite to customers and their complaints resolution times appear to be much lower, in addition to the number of complaints.

Offered QoS from above considerations and planning guidelines: Customer to be advised on the progress of complaint no later than 1 hour from the instant complaint was lodged over the telephone. In the case of written complaint, acknowledgement may be given by telephone within an hour of receipt of the complaint or by letter by return of post.

Every complaint shall be given a unique reference number (e.g. customer's telephone number) and customers advised of this. The complaint shall be logged in a computer and the 'owner' of this complaint shall also be advised to the customer. The complaint shall be retrievable by anyone dealing with the resolution of complaints.

All staff dealing with customer complainants shall be given adequate training to deal with complaints made over the phone. The training shall include an adequate knowledge of the business and the general types of complaints and their solutions. The complaint should, as far as possible, be resolved immediately; if this is not possible the complaint shall be advised of what actions will be taken. The customer should also be advised of the timescale for the completion of remedial steps and should be contacted at the end of this period.

17.3.10 Time to establish a connection

Customers' requirements: Indication of progress in connection set-up (ring tone, engaged tone, network-congestion tone, number unobtainable) in 1 second or less, from the instant the last digit is keyed, for 95% of the customer population.

Current performance level: Good for the digital network. For the analogue network delays up to 13 seconds for the longest national calls and a maximum of 35 seconds for international calls.

Considerations:

- benchmark: around 1 second for national and international calls for a digital network and up to 40 seconds for an analogue international connection;
- complaints profile: too long to make international connections to some overseas destinations;
- service-level agreements: no pressing needs;
- soft issues: customer's expectation that a dominant service provider should be able to provide quick call set-up times to any part of the world;
- strategic considerations: due to up-to-date equipment of competitors and selection of favourable international routes they are able to provide short call set-up times.

Offered QoS from above considerations and planning guidelines: The service provider to inform the customers of the plan to modernise the network by the replacement of analogue by digital transmission and switching equipment. Also to advise them that when the network is digitalised the call set-up time will meet their requirements. It should also be pointed out that nothing much can be done until the analogue network is modernised.

17.3.11 Time for resolving precontract activities

Customers' requirements: Maximum of 2 hours to resolve all precontract matters, for 95% of the customers.

Current performance level: 4 hours to deal with 95% of customers' precontract activities.

Considerations:

- benchmark: a few minutes;
- complaints profile: the main theme of the complaints is that the current time taken is too long;
- service-level agreements: not applicable, as SLA becomes valid only 'after contract' and not before;
- soft issues: poor image for the service provider;
- strategic considerations: competitors' offer much quicker response times and their average times are of the order of 1 hour or less to cover all pre-contract formalities.

Offered QoS from above considerations and planning guidelines: Target QoS set to 2 hours by the end of the following year. Customer-front staff to be trained to be versatile in all precontract activities and to be given necessary resources to complete all precontract matters in a maximum of 2 hours.

17.4 Specification of monitoring systems and determination of achieved performance

17.4.1 General

The principal discussion on monitoring systems is in Chapter 7. In this chapter the monitoring systems for the primary benefit of the network provider or the service provider are not considered. Instead, the essentials of monitoring systems to estimate the delivered performance for the benefit of the customer are considered. Monitoring systems for the benefit of customers are normally only required for the performance of parameters considered important by them. The regulator may specify these parameters based on judgement of what is important for the customers. For the purpose of this illustration, it is assumed that all 10 parameters discussed in the two previous sections need to be monitored. In the example given in Sections 17.4.2–17.4.8, the principal system requirements are given for the service provider or the network provider for the computation of delivered end-to-end performance.

17.4.2 Time for provision

The following would constitute the principal components for the calculation of delivered performance of time for provision of service by the service provider:

(a) record of time of effective contract;
(b) record of time at which provision of telephony service was complete and ready for use by the customer;
(c) sample size on which information (a) and (b) are to be kept: 100%;

(*d*) frequency of measurement: continuous, for all requests for provision of service.

From the above information the average time, standard deviation and the time for providing to 95% of the customers may be calculated.

17.4.3 Time to repair

The following information needs to be recorded to determine the time for repair:

(*a*) time at the instant a fault was reported by the customer;
(*b*) time at the instant customer was notified the fault has been repaired and the service was ready for reuse. If the customer was unavailable to be advised, the time when repair was completed would be taken for the estimation of time to repair.

From these two instants of time the time to repair can be calculated:

(*c*) sample size on which information (*a*) and (*b*) are to be kept: 100%;
(*d*) frequency of measurement: continuous, for all fault reports.

From the above information the average time, standard deviation and the repair time for 95% of the repairs may be calculated.

17.4.4 Accuracy of bills

In addition to a sophisticated and accurate charging system usually associated with the switching equipment, there is a need for a means to provide confidence that the charging mechanism can be trusted to function with the required accuracy. This is provided by a process which describes all the procedures that must be taken by the provider to ensure that the specified accuracy of the charging equipment has been maintained. The process will basically consist of checks on the charging equipment. The procedures for checking any deviance from specified performance must comply with the requirements of the *BS EN ISO 9000* [1] series of standards. An audit trail based on *ISO 10011* [2], as briefly described in Section 7.4.4, will ensure that the processes in place are being adequately followed. These procedures ought to satisfy the customers that reasonable steps have been taken to ensure that the specified accuracy of the charging equipment is being monitored.

17.4.5 Network-related parameters

The network-related parameters are availability of network resource to establish a connection, call quality, misrouted calls and availability of network resources to maintain the connection for the intended duration of the call.

The monitoring of network-related parameters is discussed in Chapters 7 and 10.

17.4.6 Time for resolution of complaints

The following information needs to be recorded to determine the time for the resolution of complaints made by the customer:

(a) time at the instant a complaint was reported by the customer;
(b) time at the instant customer was contacted regarding the outcome of the complaint. If the customer was unavailable, the instant at which complaint was resolved effectively would form the basis for the estimation of time for resolution of complaints. From these two instants of time, the time for the resolution of complaints can be calculated;
(c) sample size on which information (a) and (b) are to be kept: 100%;
(d) frequency of measurement: continuous, for all complaints, irrespective of mode of making complaint.

From the above information the average time, standard deviation and time for resolving 95% of the complaints may be calculated.

17.4.7 Time to establish a connection

This parameter needs to be measured only where analogue systems exist and digital and analogue systems interwork. Time to set up calls in an all-digital network is not considered a problem area for basic telephony service. The call-set-up times may be measured, for analogue networks, for both national and international connections, and for combinations of analogue and digital connections, both national and international.

Monitoring system: The type of measuring device will have the following features: it will detect the destination from the address digits; identify receipt of the last digit and commence a timer; identify the network signal when indicating the status of the called party (e.g. as called-party contacted — ringing tone, called-party engaged, called-party unobtainable) or if the network is unable to establish any of the above due to network congestion. The time to answer by the called party is not relevant, only the time taken to establish the connection.

The pattern of calling behaviour may be studied and samples taken to obtain representative call-set-up times. The sample size shall be determined by the variations on the measurements. The measurements may be carried out to individual countries. Consideration ought to be given to peak traffic periods in both the calling and called countries. Consideration should also be given to parts of the called country which may have analogue networks and other parts which may have digital networks. A representative set of connection-set-up times may then be established. These measured values are unlikely to change with time unless changes in switching equipment are carried out in either country or in a transit-switching country. The call-set-up times need to be reviewed only when changes in switching equipment have taken place.

Measuring devices carrying out the above tasks exist and few such instruments are needed to cover international and national routes. Such measurements will become superfluous in an all-digital environment.

17.4.8 Time for resolving precontract activities

The following information needs to be recorded to determine the time for resolution of all precontract activities between the customer and service provider:

(a) time when the first approach was made by the customer to the service provider;

(b) time at the instant customer was given sufficient information to enable a contract to be placed.

From these two instants, the time for precontract activities can be calculated.

(c) sample size on which information (a) and (b) are to be kept: 100%;

(d) frequency of measurement: continuous, for all queries;

(e) must survey, separately, different groups of customers, particularly those at risk from competition, e.g. major business customers.

From the above information the average time, standard deviation and the time to serve 95% of the customers may be calculated.

17.5 Customers' perception of quality

The management of customers' perception from surveys has been dealt with in Chapter 8. Figure 17.3 shows a hypothetical distribution of delivered performance on repair of faults. All repairs have been carried out in 12 hours and 95% in 6 hours.

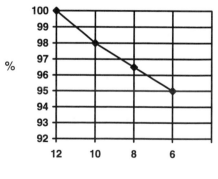

Time to repair (hours)

Figure 17.3   Delivered performance: repair time in hours against percentage of repairs carried out — a hypothetical case

If the customer survey for this parameter indicates that 95% of customers are satisfied with the perceived performance of repair, it can be concluded there is a 1:1 correspondence between their perception and their requirement as the requirement for repair time, from Section 17.2.3 was for a maximum of 6 hours for 95% of the customers. However, if the survey showed that only 80% of customers were satisfied, indicating lack of 1:1 correspondence with their perception of quality for this parameter, the following avenues have to be explored for possible explanation:

- customers' expectations have increased, leading to an expectation of better performance;
- competitors' better performance offerings have influenced the customers' perception of delivered quality; or
- business requirements, since the capture of performance requirements, may have become more stringent, leading to higher expectation of performance.

It is important for the service provider to identify accurately customers' perception of quality. Analysis of variance requires a mixture of professional skill and knowledge of the customers' business culture.

The principal factors to be borne in mind in the practical management of customers' perception of quality are:

(a) the survey has to be carried out on a service-by-service basis;
(b) customer perception has to be obtained on a parameter-by-parameter basis for comparison with the delivered performance;
(c) the parameters chosen must be those considered of important by customers; these may differ from those considered important by the service provider; and
(d) the frequency of such surveys must ideally coincide with the review of the viewpoints of the quality cycle.

17.6 Service-specific QoS requirements: personal communications

In the management of QoS, it is vital to identify the key issues of quality requirements on a service-specific basis and, if necessary, on a population-segment basis. To illustrate this, some of the service-specific requirements of personal mobile communications are listed below followed by the specific requirements of a professional user segment of the population.

Common QoS issues on the provision of mobile communications are:

(i) call-set-up times for mobile to land, land to mobile and mobile to mobile communications to be indicated;
(ii) the effect on probability of blocking due to additional network elements needs to be quantified;
(iii) scanning for a better-quality channel and its implications to the user;

(iv) effective handover to adjacent cell;
(v)   transmission quality to be good and contiguous inside and outside a building and over a large geographical area;
(vi) avoidance of fading of signals in certain terrain;
(vii) call-forwarding facility to be provided;
(vii) storage of message when personal terminals switched off;
(ix) security of conversation, i.e. no overhearing; and
(x)   unprompted release of calls.

The professional-user radio systems or private mobile radio (PMR), whose function is to provide what may be considered a closed user group, may have the requirements listed in Table 17.3 which are different to the average mobile user.

Table 17.3   Requirements of professional users of private mobile radio [3]

Service/ characteristic	Professional user	Mobile telephone user
Ability to operate in a group, with all members of the group hearing all conversations	Yes	Generally not required – limited capabilities can be provided by conferencing
Capabilities to allow dispatcher to control users	Yes	No
Call-handling supplementary services	Tailored to each type of user and many PMR supplementary services (SS) unlike telephone SS	Standard across entire system and similar to telephone SS
Press-to-talk-type access to other users and the despatcher, i.e. no need to dial	Essential	Generally not required
Near-zero call-set-up times	Essential for police etc.	No
Scale of system	From single site to nationwide	Regional or national coverage
Very high capacity	Usually not required	Essential
Frequency planning	Often no frequency co-ordination with other nearby users	Frequencies co-ordinated across entire system
System provision	Often owned and operated by user, although there is significant use of shared (PAMR) systems	Services provided by system provider
Payment for the service	Pays for provision; use is 'free'	Pays for how much of the service is used

Identification of key QoS issues on a service-specific and, where necessary, on a population-segment basis of all services offered by the service provider will provide the basis on which further management actions are to be taken in the management of QoS.

17.7 Common requirements on QoS

Some quality issues are not necessarily service specific and could cover more than one service. Examples of these are:

- 'geographical hot spots' of poor quality;
- network congestion due to special events, e.g. voting by telephone on a television programme;
- customers' repeat-call attempts at times of network congestion causing further congestion.

17.7.1 Geographical hot spots

Management of hot spots (also known as 'tail management' due to the appearance of a 'tail' if repair times are plotted on a graph) in the management of any quality parameter is matter of concern to any service provider. This term is applied when the incidence of a fault is significantly higher in one geographical area than in the rest of the country. The effect of the hot spot may be masked if the national average is given, but is highlighted if regional performances are indicated. Resolution of any form of hot spot requires systematic examination of the likely causes before remedy can be achieved. For a description of tail management, see the contribution by Smith and Tidswell [4].

17.7.2 Network congestion due to special events

Certain television programmes are followed by requests to the public to vote on the contents of the programme or the personalities involved in it. Usually, a time limit is given and for this period there is a larger-than-normal requirement of network resources to switch through calls. This often leads to network congestion (see also Section 10.6).

In the management of this type of congestion, the service provider resorts to various solutions; one is to ensure that such programmes take place at periods which are not the busiest and a second is to provide additional switching equipment. Call gapping (where every *n*th call is not allowed to go further into the network) is an option which may be used to prevent undue congestion. The provider's solution will depend upon the resources available at their disposal.

17.7.3 Congestion due to repeat last call

During busy hours, when network congestion is present, a certain number of call attempts would be unsuccessful. The customer receives a network-congestion indication and would be inclined to repeat the call attempt soon. This introduces additional congestion and can produce further degradation of switch performance due to the higher number of call attempts. A study was carried out by one of the authors to estimate the degradation of exchange switching equipment at different levels of overload. Assumptions were made on the likely number of repeat attempts by a user, the time interval in which these attempts are likely to take place and the proportion of users that will make repeat attempts. For the increased call attempts the degradation of switch performance may be estimated from the manufacturer's performance characteristic for the switch.

The solution to this type of network congestion is additional switching capacity. The actual addition of capacity would depend upon the calling pattern at the switch.

17.8 Conclusions

What appears to be a rigorous approach, should, in time, prove to be a straightforward exercise and lead to a better management of quality than the 'fire-brigade' approach. Merely reacting to emergencies is a sign of poor planning. In a competitive environment, this approach could prove to be disastrous in the long run. Such an approach satisfies neither the customer nor the service provider.

17.9 References

1   ISO 9000: 'Quality management and quality assurance standards'
     ISO 9001: 'Quality systems – model for quality assurance in design, development, production, installation and servicing' (International Standards Organisation, 1994)
2   ISO 10011: 'Guidelines for auditing quality systems. Part 1: Auditing; Part 2: Qualification criteria for quality system auditors; Part 3: Management of audit programs' (International Standards Organisation, 1993)
3   MACARIO, R. C. V. (Ed.): 'Modern personal radio systems' (IEE, 1996), p.80
4   SMITH, D., and TIDSWELL, D.: 'Tail management', *Bri. Telecom. Eng.*, 1995, **14**, pp. 115–121

Chapter 18
Quality of service: the future

18.1 Introduction

The telecommunications industry is undergoing an unprecedented and accelerating rate of change and increasing rapidly in complexity, with ever greater competitive and financial pressures. In such an environment of turbulence and uncertainty, it is difficult to speculate on the future with any certainty. Bob Allen, chief executive officer of AT&T, in a speech in 1994 at the Detroit Economic Club said 'The exact shape of things to come is as unknowable today as it was in the time of primordial ooze, and some of our visions may prove as fanciful as pigs with wings'. Nevertheless, it is important to form a view of the potential impact on QoS of future industry developments in order that evolution takes account of future customer requirements and provides maximum competitive differentiation.

The key issues to be addressed are likely to be:

(a) To survive and prosper in today's challenging environment, network operators are having to seek new markets overseas. The increasing globalisation of industry, characterised by the growth in multinational companies, provides an attractive source of revenue which is being sought by a small number of large operators. This is precipitating the erosion of the traditional bilateral 'correspondent' relationships traditionally used for international telecommunications services. However, the delivery of sophisticated global services requires alliances between operators which fragment the responsibility for end-to-end QoS. Additionally, there is a need to understand the impact on perceived QoS of the different country cultures in which a global operator trades.

(b) The evolution to the, so called, 'information society', together with the pace of technological change, is giving rise to new customer requirements and means of delivering multimedia[†] information that will erode the traditional telecommunications infrastructure and its value chain. The convergence of the telecommunications, information-technology (IT), media and entertainment industries is creating an information market with a plethora of collaborating and competing players. The QoS dimensions for new innovative information services have no precedents from which to extrapolate, e.g. when shopping from

† Multimedia: information with a mix of graphics, text, moving pictures and sound.

the television screen, what are the customer requirements on true representation of colour, contour and 'feel' of the goods advertised ?

Other questions to be addressed include the role of standards and who will formulate standards for this diverse mix of industries which will provide tomorrow's information services. What is the role of the regulator in the future quality issues? Should it be reactive or proactive? Who will be the regulator? What role could the user groups play in bringing about progress which will benefit all parties? What about the network provider and the service provider? Should their loyalty be only towards the shareholders (where these exist) or to the government? Should they be an independent supplier of carriers? Do they have a societal role to play or only that of a provider of secure networks?

In the following sections the key issue of the information society is discussed. This is likely to be the most dramatic change which network operators and service providers are likely to experience and it will pose interesting problems of how to identify and satisfy customer QoS requirements.

18.2 Impact of the information age

In the 1950s, Peter Druker, widely acknowledged as one of the founders of modern management theory, predicted a society shaped and altered by information. In describing the impact of such changes, Druker coined the term 'knowledge worker' to describe those who use intellectual skills to process information and generate wealth [1]. In the early 1990s, the prediction appeared to be becoming true when the industrial age gave way to the information age and, for the first time, companies in the developed world spent more on computing and communications than on industrial, mining and farming machinery [2].

Vice President Al Gore of the USA articulated his vision of the future as 'an environment that stimulates a private system of free flowing information conduits ... innovative appliances and products giving individuals and public institutions the best possible opportunity to be both information customers and providers' [3]. Gore's vision of the 'information super highway' became policy with the creation of the National Information Infrastructure (NII) initiative and the establishment of the Information Infrastructure Task Force and an associated advisory council.

In Europe, a report produced by the Bangemann Committee in June 1994 [4] examined the measures to be taken by the European Union in relation to the emerging information society. Subsequently, the European Telecommunications Standards Institute (ETSI) produced a detailed report on the likely environment, industry structure and network infrastructure [5]. At the same time, the European Institute for Research and Strategic Studies in Telecommunications (EURESCOM) produced a joint view of the European public telecommunications operators of the information society and its needs [6].

In the UK, the Office of Telecommunications (OFTEL), the regulatory body, issued a consultative document 'Beyond the telephone, the television and the PC' [7]. This dealt with regulatory issues related to broadband switched mass-market services delivered by telecommunications systems. The document pursues four complementary goals, namely:

- to remove regulatory uncertainty, so as to facilitate investment and innovation by all potential players;
- not to undermine the existing regulatory framework;
- to facilitate continuing investment in competing infrastructures;
- to set out a framework for the regulation of the broadband, switched mass-market (BSM) services of the future.

In February 1996, the UK Department of Trade and Industry (DTI) launched the Information Society Initiative (ISI). Set to run until the year 2000, the initiative comprises a comprehensive range of activities – from providing information and guidance, to funding model projects and running research programmes.

Japan has also recognised the strategic importance of building an 'info-communications infrastructure' with the issue, in May 1994, of a Government report titled 'Reforms towards the intellectual creative society of the 21st century' [8]. Subsequently, an 'advanced info-communications society promotion headquarters' was established in the Cabinet for the integration of promotion measures and co-operation in world-wide efforts towards the GII [5].

The Global Information Infrastructure (GII) initiative was launched by Al Gore at the first World Telecommunications Development Conference in March 1994. This was consolidated at a G7 conference, in January 1995 in Brussels, attended by heads of Government of the Group of Seven (G7) industrialised countries. The conference discussed the translation of policy into action and the need for close co-operation between the world's three major trading blocs, i.e. North America, the EU and Japan.

The above evidence suggests that the information infrastructure will happen, but there is great uncertainty about what services will be offered and by whom, together with customer usage and QoS expectations. The nearest existing model is the Internet. Widely viewed as the first step towards the creation of the GII, the Internet evolved from the US Department of Defense 1960s project to interconnect facilities undertaking research. By 1973, this ARPAnet had become a network interconnecting islands of computer resource used by the research community and academia across America. By the late 1970s the network had spread outside of America and it is now a global network experiencing an ever-increasing year-on-year growth rate. There is no single controlling body for the Internet, except for the allocation of addresses and the interconnection of the various domains. The network protocols evolve from a consensus-based process carried out by the Internet Engineering Task Force (IETF). Resulting from the academic background of the Internet, free access to information has traditionally been provided. But the means of access was not user friendly, so usage was restricted to the computer-literate fraternity. However, the recent introduction of user-friendly browser applications, e.g. Mosaic and Netscape, has resulted it a growing variety of users and the beginning of electronic trading.

The Internet is not a managed network. It uses its best efforts to pass information from source to destination and the QoS depends on how many users the network is serving at a particular time. If there is traffic congestion, the delays increase. It can carry real-time voice and video traffic but cannot guarantee comprehensible delivery. Charging is normally by a flat fee to Internet Service Providers (ISP) with unlimited usage and the charge of a local call to the ISP server, but there are no service guarantees. The QoS leaves much to be desired, but there is no alternative. Nevertheless, experience of the Internet gives some useful pointers about QoS aspects which will be important for the future information infrastructure.

18.3 Services of the information infrastructure

The information infrastructure must support the requirements of users who will dictate which services are provided and the QoS. However, the convergence of IT, information and entertainment has the potential to provide a rich variety of innovative multimedia services, as illustrated in Figure 18.1. These will be made available to end users by the use of digital techniques which facilitate the economical processing, storage and transport of information. However, it is likely that the current perception of future information services will, given the pace of change, be substantially

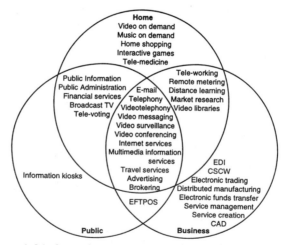

Figure 18.1   Potential information-age services

EFTPOS electronic funds transfer at point of sale
CSCW computer-supported co-operative working
CAD computer-aided design
EDI electronic data interchange

erroneous, and many new services will emerge that cannot be foreseen from the experience of today's environment.

The market for such services can be coarsely segmented into home, business and public. But, as illustrated in the Figure 18.1, there may be a considerable overlap between the services and markets. This may cause QoS conflicts if the requirements of customers in each of the segments are different. However, in practice, the segmentation of the information market will be of much finer granularity, perhaps approaching a single-user segmentation, with significant differences in QoS requirements and perceptions. As demonstrated by the Internet, information has no national boundaries. Hence the information market will be global and the difficulties in understanding customer requirements will be compounded by cultural differences in what is perceived to be an acceptable QoS.

Although end users will drive the requirements for innovative information services, it should be appreciated that the delivery of such services will require services·to be provided between the various players within the overall information industry, e.g. transport of information between the players by the network operator.

Generic infrastructure requirements for end-user services identified by ETSI [5] include:

(a) the ability to capture and store content at an affordable cost;
(b) user interfaces and navigation methods which work for the user on any type of terminal;
(c) privacy, authentication and data security;
(d) a simple, efficient and effective billing system;
(e) broadband capacity sufficient for wide access to a large number of coexisting service providers;
(f) real-time access, low access time and minimum risk of loss of connection for some services.

18.4 Roles and relationships in the information industry

It is likely that the new information marketplace will create many new players, some of which will challenge the traditional role of telecommunications operators as providers of the communications infrastructure. The European Telecommunications Standards Institute (ETSI) [5] views the role and responsibilities of the various players in terms of a primary information-value chain which starts with timely raw information, which may lack corroboration and context; and ends with information presented to users in a manner that can be easily understood and used. Each of the structural roles in the primary chain adds value and sells to the next role in the chain. The concept of the value chain was introduced by Michael Porter in his book 'Competitive advantage' [9]. Porter has used it to 'disaggregate' a company into its strategically relevant costs in order to understand, for example, the behaviour of costs. However, the term has been more widely used to describe an industry structure into which a single enterprise fits (Porter describes the wider structure as the 'value system'). An example of the application of the value chain to the information industry is the film industry where the allocation of box-office revenue to the various elements in the chain is broadly as follows [10]:

Cinema operator	19
Distributor	24
Advertising	20
Production	31
Sundries	6
Box office revenue	100%

The current ETSI view of the structure roles in the information-industry value chain, and the activities of the participants, is shown in Table 18.1

The Information value chain (Figure 18.2) starts with raw information which has the key attribute of timeliness but is not necessarily saleable if it lacks corroboration and context. Next comes processing to give it marketable contextual validity. It may then be stored pending retrieval by the broker or end user. Information has no value unless it can be found; thus the search part of the chain is particularly important and must minimise the efforts of the end user. The structure roles in the value chain are supported by infrastructure roles such as the provision of communication facilities to pass information between structural roles. Also supporting the chain are the

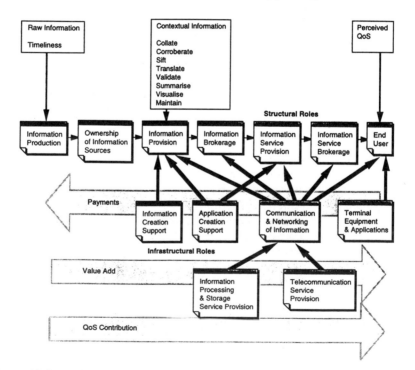

Figure 18.2   Information-value chain [5]

various applications which facilitate manipulation of the information, e.g. authoring tools for information creation and processing.

Table 18.1   Information actors [5]

Roles	Activities
Source-information ownership	Selling rights of recordable events Using intellectual property rights
Information production	Creation, capture and production of information Creation and production of information services and Applications of information services
Information provision	Creation, capture and production of information Presenting and advertising to information brokers and service providers Selling information to service providers Delivering information to service providers
Management of information provision	Accounting, billing, revenue sharing Auditing Providing access rights Assuring against illegal information movements
Information bokerage	Registration of information providers and their offerings Presenting, advertising and trading information Provision of navigation for information-service providers to information providers
Information-service provision	Searching, requesting and purchasing information from information providers Presenting and advertising to information-service brokers, end users and other information-service providers Searching, requesting and purchasing information services from other information service-providers Selling information to end users and other information-service providers Delivering information services to end users and other information-service providers
Management of information-service provision	Accounting, billing, revenue sharing Auditing Providing access rights Assuring against illegal information movements
Information-service brokerage	Registration of information-service providers and their offerings Presenting, advertising and trading information services from information-service providers to end users Management of the access rights of end users to information-service providers Provision of navigation for end users to information-service providers
End user	Consumption of services and applications Combining information-industry services with other services Defining and specifying requirements
Regulator	Development of regulation Encouraging production of standards for competitive supply Licensing of players

A simplified view of the information-value chain is presented in Figure 18.3. It comprises four elements for content provision, service provision, distribution and consumer equipment. There are many companies occupying parts of this value chain, but a number of the large companies in the entertainment, media, IT and telecommunications sectors are beginning to position themselves throughout the chain (by alliances, joint ventures, mergers and acquisitions) to improve their competitive position in the information market. For example:

(a) BT (telecoms): Traditionally operating in the service-distribution, service-provision and consumer-equipment sectors, BT has concentrated in gaining global distribution reach with its 'Concert' joint venture with MCI in the USA and numerous alliances in many countries to gain customer access. However, it is moving into the content-creation sector (e.g. 'yellow pages'), and the content provision with a number of online services focused on specific market sectors (e.g. education, health, construction etc). It is also trialling video games and interactive video services (e.g. video on demand and video shopping etc.) and, via Concert, offers global Internet access with Concert InternetPlus.

(b) Microsoft (IT): The company dominates the operating-system, middleware and applications-software market, but has been rapidly moving into all sectors of the value chain. Examples include its joint venture with NBC for 24-hour broadcast news and online news service (MSNBC online), its purchase of eShop for online shopping and its deal with AMEX to provide an Internet corporate travel service. Its alliances include AT&T, Nintendo, Paramount, DirectTV, various network operators, CompuServe, AOL, TCI and many others.

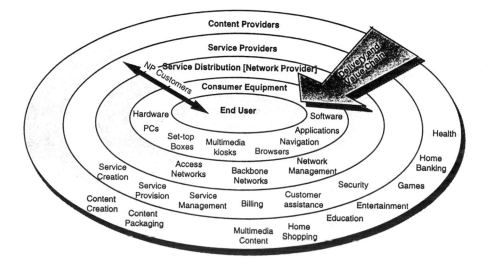

Figure 18.3   Information-value chain — simplified view

(c) Disney (entertainment): Best known for its films, it is moving into other areas of the value chain. Examples include its acquisition of Capital Cities/ABC for distribution, its theme parks and resorts, Disney Interactive CD-ROM production, Disney Online, 500 Disney stores, joint ventures (JVs) with various network operators, investment in Miramax for sensor technology etc.

(d) News Corporation Ltd (media): Rupert Murdoch's empire is best known for newspapers, publishing and satellite TV, but is now extending its world-wide distribution by satellite-broadcasting ventures, acquisitions and joint ventures (JVs) in USA, South America, China and Asia and an alliance with MCI and cable TV in Australia. It runs an online news service and is moving into multimedia production (video and CD-ROM) with its purchase of Shomega.

The traditional telecommunications network operator may not be best placed to respond to the needs of the emerging information society. The value and payback for transport and delivery of information is likely to be a small proportion of the total value chain. Also, the convergence of content, computing and telecommunications will give rise to more and different competitors who are fleeter of foot and more commercially adept. Note, also, that the network operator now occupies only part of the value chain and will only get a proportion of the revenue from end users, depending on what value is added.

The share of the revenue obtained from the end customers may also depend on the number of service providers that can be attracted to connect to the operators' network. Hence, the network operator will need to consider both service providers and end users as customers, giving both the QoS that is required and that will differentiate it from competing network providers (distributors).

18.5 Network architecture

It is widely accepted, by the telecommunications community, that the bit-transport layer of the broadband ISDN (B-ISDN) which will support the information infrastructure will be based on an asynchronous-transfer-mode (ATM) switched network with a synchronous-digital-hierarchy (SDH) transmission-bearer network. The IETF of the computer industry favours a network based on Internet-protocol (IP) routers, as used by the Internet. Given the investment in both technologies, it is unlikely that one will replace the other. The most likely outcome is that the two technologies will converge to complement each other, as proposed by the IETF 'IP over ATM' working group.

Traditionally, the telecommunications network comprises switching nodes interconnected by links with well defined interfaces and signalling protocols.

There are user-network interfaces (UNIs) between the network and customer-terminal equipment, also network-node interfaces (NNIs) in the network and between networks. End-to-end quality of service is heavily influenced by the summation of the performance of individual nodes and links in a connection. In addition, there are application-programming interfaces (APIs) to the control software to enable changes and upgrades to be made, but these are usually proprietary and closed.

Traditionally, network operators provide services from applications embedded in their networks. This is a closed environment, the only open interfaces being between the networks of operators at the transport level. However, the upcoming information environment will require open APIs between all structural roles to support application development and access to the various storage, processing and transport facilities.

A more appropriate approach to the future structure may be to consider the functional blocks required for the information infrastructure. The functional approach has been developed by the Digital Audio-Visual Council (DAVIC) [20] which has defined about 250 functions and their open interfaces. Combinations of these functions, when put together in a specific way, can provide individual services. However, there are 10 core functional groups which are basic to the needs of information services.

Applications will invoke the core functions as required, or replace or extend them with more specific functions (which may in turn invoke core functions). The core functions are:

(a) Bit transport provides the physical and logical links for the required bandwidth between the required points to be connected.

(b) Session control controls the bit-transport functions, calling on them to establish or change a logical connection and determining the data rate and protocols to be used.

(c) Access control provides facilities to authenticate the user and verify access rights to the network, specific applications and related content, goods and services. It also provides verification of credit and payment.

(d) Navigation, programme selection and choice enables the user to find and choose application or content, probably by the use of hierarchical menus.

(e) Application launch provides the facilities to run an application, obtaining and loading any necessary code (if not resident).

(f) Media-synchronisation links provide links between objects, i.e. sound segments, subtitles, still and moving images, and applications to achieve a multimedia presentation.

(g) Application control provides control of an application's behaviour, e.g. pause, rewind etc., or content options/interactions.

(h) Presentation control is for control of delivery and display of multimedia information, e.g. subtitle activation, choice of language etc.

(i) Usage data means the collection, storage and supply of data relating to users' consumption of material, resources and applications, for such things as payment, marketing and resource utilisation etc.

(j) User profile stores and utilises information about individual users and their behaviour to control access, assist navigation and bill correctly for services received.

Applying these functions to a broad customer-centric model of the information industry, i.e. customers, service providers, content providers and network operators, gives the possible mapping shown in Figure 18.4. Note that the drive towards open interfaces and the experience of Internet indicate that only bit transport may fall into the domain of the network operator. Distributed-computing concepts such as JAVA and Active-X give a wide range of options for the location of functionality within the industry model.

The model with its functional building blocks begins to resemble that used by the IT industry. There, for example, IBM, Compac etc. provide the hardware platforms which support a common (open) operating system (e.g. DOS); Microsoft Windows offers an open middleware platform which can be used by applications developers to efficiently develop, deliver and simplify usage efficiently by a common format and functions such as 'open', 'print', 'save', 'cut', 'paste' etc. It could be considered that the bit-transport network with its basic functions such as call control, redirect, calling-line identity etc. is the equivalent to the hardware/DOS platform. The telecommunications middleware platform contains value-add functions such as billing, authentica-tion, network management etc. with open interfaces to the service-control point (SCP) and associated service-creation and management environment. Hence, it is this middleware component that is likely to be the battleground between the telecommunications and IT industries, since the service functionality can be provided either from within the network or at the periphery.

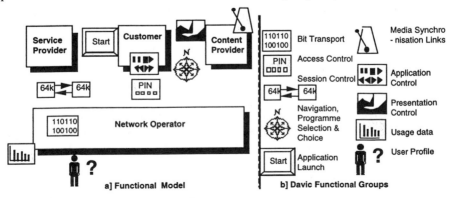

Figure 18.4   Information-infrastructure functionality

Network control can be considered as two layers, namely basic call and bandwidth control, and the middle ground for many of the DAVIC functional groups such as navigation, access control, etc. which may be located in the network or at its periphery. Intelligent CPE can interact directly with services which interface with content. These functional elements are shown in Figure 18.5 as a layered architecture with open application-programming interfaces (APIs) between each layer which allow any network operator or service provider to use any combination of layers in the stack. This architecture will be the same for national and global networks

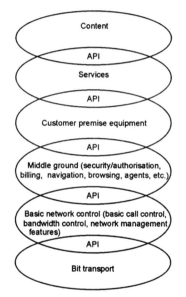

Figure 18.5   Future network architecture

18.6 Quality of service in the information industry

18.6.1 User requirements

With so many roles in the value chain, interoperability is likely to have a major impact on the QoS perceived by end users. There will need to be a transparent transport path between the server network of the service provider and the end user via the various delivery networks. Clearly, there needs to be sufficient bandwidth in the various network segments of the end-to-end connection; the transmission performance of the network segments must be compatible with the needs of the service, and the transfer mode (packet- or circuit-switched) and transport formats of the segments must also be compatible. This technical interworking relies on the development of appropriate interface and protocol standards. In a similar manner, there must be compatibility between the elements involved in the control of the particular service and this is effected by the various API and middleware interfaces/protocols. There must also be interoperability between the various naming and addressing schemes involved in the navigation from the end user to the required information. Lastly, interoperability of the information formats can have a major impact on QoS. This is already a serious problem for word processing with incompatibility between the various packages and even between different versions of the same package.

The performance requirements for ATM-based bit transport are being defined by the standardisation fora and are discussed in Chapter 9. Of the traditional transport parameters, traffic congestion, which is currently well controlled by the use of such techniques as automatic and dynamic traffic routeing together with traffic management, may well prove difficult to control. The current 64 kbit/s telephony-based network will be replaced by a multibit-rate broadband network where per-call transport requirements can vary from a few kbit/s to many Mbit/s and call-holding time can vary from a few minutes for a telephone call to well over one hour for a video-on-demand (VOD) session. Thus one 1.5 hour VOD session delivered at 2 Mbit/s is equivalent to 900 telephone calls at 64 kbit/s each of 3 min duration. Hence demand on the network will be difficult to forecast. There is nothing to extrapolate from, and the dimensioning of the network could result in large underprovision for quite small errors in broadband forecasts. Additionally, surges of broadband traffic will be difficult to control, since they will consume vastly more network resources than the equivalent 64 kbit/s surges experienced today.

In the information age, the QoS perceived by end users may well be impacted by all the role in the value chain, from information creation to delivery. QoS will not only be seen in terms of the 'technical' quality but will be heavily influenced by the usefulness of the information, the ease and speed of finding it and the simplicity of comprehending it.

To illustrate the range of quality issues to be addressed, a list identified by a recent study by the European Commission [21] is given in Appendix 4. Study of the requirements shows that users can be very demanding on quality matters. The list also shows that a wide range of issues are to be addressed for lasting satisfaction among users of electronic mail (e-mail), a relatively simple information service. The list is intended as a guide to future service providers, network providers and regulators, for identification of future QoS issues.

For information services, a clue to end-user requirements may be obtained from what exists for the industries that are merging into the information industry (i.e. telecommunications, information, computing, commerce and entertainment), and how these industries are separately regulated. For example:

(i) Many new services will revolve around publishing and advertising industries which have their own quality issues and codes of practice.
(ii) The rise in electronic commerce will need to adapt the laws and practices currently formulated for paper-based processes.
(iii) The requirements for television entertainment are well understood.

Other QoS aspects include:

(a) Multimedia: The user requirements for multimedia communications services are particularly onerous for desktop collaborative working where users could experience the combined effect of the full range of media. Useful work is being carried out by the Multimedia Communications Forum [22] to determine user QoS requirements and the complementary network- and terminal-performance parameters. This is reported in Chapter 3.

(b) User interface: Basic user requirements encompass the ease with which they can obtain timely information and its usefulness. However, historically, the barrier between end users and the benefits of IT has been, for all but the technophile, the unfriendly user interface. This is a major quality and usage constraint. Human factors will play an increasingly important role in the design of user-friendly terminal equipment which will be a critical factor in the user choice of supplier [11] Aspects contributing to ease of use include:

- voice recognition to overcome poor typing skills;
- touch sensitive screens to improve the point-and-click functions of a mouse;
- visualisation of information [12,13] to assist in converting raw data into meaningful information, an increasing necessity in a future era of information overload;
- enhanced navigation to improve the ease of finding information [14,15]. To assist the user to find the information required, electronic agents will be employed as shown in Figure 18.6. The customer agent will negotiate with information provider agents to find the required information at the

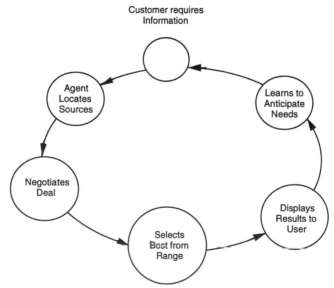

Figure 18.6   Agents

right price, e.g. the cheapest option for a holiday at a specified resort and date. There may well be negotiation with a network agent to transport information to the users terminal. Over time the user's agent will learn its owner's habits to anticipate needs [16].

(c) Contracts: Current Internet commerce is similar to 'browsing' through newspaper and magazine advertisements and purchasing goods. However, many UK newspapers and magazines are members of the Mail Order Protection Scheme, which gives the consumer some redress if the supplier fails to discharge its obligations, but there appears to be no equivalent for the Internet.

Since there can be no guarantee that electronically communicated documents have not been altered by the recipient or a third party during transit, there is no indisputable audit if conflicts occur. Business-to-business electronic commerce is typically underpinned by paper-based contracts agreed in advance. There are some standards, the so-called, interchange agreements for electronic trading produced by bodies such as the UN, EU, International Chamber of Commerce [17] and the Bolero Association. The latter is producing E-terms (electronic terms and conditions) which can be used generally for electronic trading on a global basis [18].

(d) Security: Security is a major issue for electronic trading, and this will put heavy responsibility on the 'carriers' to transport information without corruption. Also, suppliers and consumers will need to be convinced that spending money online is at least as secure as normal cheque and credit-card transactions. A number of solutions are being considered, such as, key authentication, where messages are decrypted using the originator's public 'key' and that of a 'trusted third party'. This proves that the message came from the originator, since it requires the assistance of the Trusted Third Party. Timestamps are the equivalent of the 'recorded delivery' of the postal service and are necessary to authenticate delivery of a message.

(e) Censorship and content control: Content control is currently a major issue with the Internet. In the USA, the Communications Decency Act of 1996 (an element of the 1996 Telecommunications Act) creates criminal penalties for the 'knowing' transmission of material considered to be 'indecent to minors'. It puts the responsibility of content control onto service providers with the possibility of prosecution if their customers contravene the Act [19]. Not surprisingly, the act has been challenged legally by the free-speech lobby.

(f) Intellectual property rights and copyright: These are fundamental to the development of information services. Without them, content creators may not wish to make their works electronically available. The World Intellectual Property Organisation (WIPO) is currently formulating proposals for a protocol which extends the Berne Convention on copyright to recognise developments in electronic media.

18.6.2 Service-provider needs

It is a given that both the end user and service provider will require guarantees that contracted, on-demand bandwidth will be met within contractually agreed performance parameters, with legal recourse if these service-level agreements are not met. However, for a service provider to require to be connected to the network of one operator rather than another, it will require:

(a) the operator's customer base to contain a reasonable proportion of its target market;

(b) ease of interconnection; and

(c) the possibility of outsourcing some of its service-management activities, such as microbilling.

Ease of interconection will require application interfaces to the network that will allow applications to be run using network 'middleware', in much the same manner as Microsoft Windows offers a middleware platform which application developers can use effectively to develop and deliver their applications, e.g. offering generic functions such as 'print', 'save' 'cut', 'paste' etc. with open interfaces which can be used by any application.

The cost of delivering information over the information infrastructure is likely to be small and will provide a problem to service providers in billing for low-priced transactions for large volumes of customers. The network operator, however, operates billing systems which deal with huge volumes of small transactions, i.e. individual calls from a customer base of millions. Since the customers are the same for the network operator as the service provider, it appears logical that this microbilling should be carried out by the network operator on behalf of the service provider. However, it will impose onerous billing-quality requirements on the network operator.

18.7 Standards

The merging of a number of industries to meet the needs of the information society has resulted in a large number of disparate fora contributing to formally agreed (*de jure*) standards necessary for information services. In addition, the rapid pace of development may well spawn market-driven industry/company-based *de facto* standards. Such standards are likely to impact on the QoS delivered to end users.

The web of standards fora for the information infrastructure is illustrated in Figure 18.7. Although collaboration between the various bodies is improving, it is far from satisfactory and there appears to be no common consensus on customer requirements.

Within Europe, two bodies have been established to bring more coherence to the standardisation necessary for the European Information Infrastructure (Figure 18.8). They are:

(*a*) High Level Strategy Group (HLSG) for information- and communications-technologies (ICT) standards. This was set up in 1995 to 'identify and facilitate timely development of missing, critical ICT standards/specifications' necessary for the implementation of the information society in Europe. The four trade associations European Telecommunications Network Operators (ETNO), European Telecommunications and Professional Electronics Industry (ECTEL), European Association of Manufacturers of Business Machines and Information Technology Industry (EUROBIT) and European Association of Consumer Electronics Manufacturers (EACEM) are represented. A number of projects have been carried out which identify barriers to be overcome by the standardisation bodies; also the legal and regulatory requirements necessary for ICT applications to be launched in the market.

(*b*) ICT Standards Board (ICTSB). This was established in 1995 and has 15 standardisation bodies as members. Its role is to analyse the requirements for the European Information Infrastructure (EII), translate them into standardisation programmes and allocate the work to the most appropriate forum.

(*c*) European Programme on the Information Infrastructure Co-ordination Group (EPIC). This was set up as a EII standardisation-management committee and became the technical resource arm of the ICTSB in 1996. About 30 projects were set up, divided into four groups or 'highways', namely:

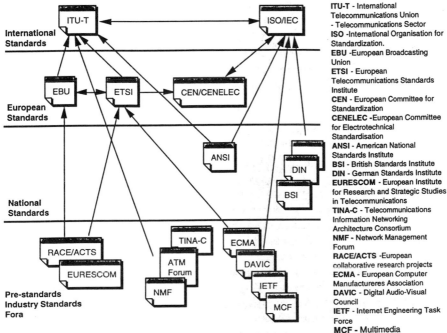

ITU-T - International Telecommunications Union - Telecommunications Sector
ISO -International Organisation for Standardization.
EBU -European Broadcasting Union
ETSI - European Telecommunications Standards Institute
CEN - European Committee for Standardization
CENELEC -European Committee for Electrotechnical Standardisation
ANSI - American National Standards Institute
BSI - British Standards Institute
DIN - German Standards Institute
EURESCOM - European Institute for Research and Strategic Studies in Telecommunications
TINA-C - Telecommunications Information Networking Architecture Consortium
NMF - Network Management Forum
RACE/ACTS -European collaborative research projects
ECMA - European Computer Manufactureres Association
DAVIC - Digital Audio-Visual Council
IETF - Internet Engineering Task Force
MCF - Multimedia Communications Forum

Figure 18.7   Standards for the information infrastructure [23]

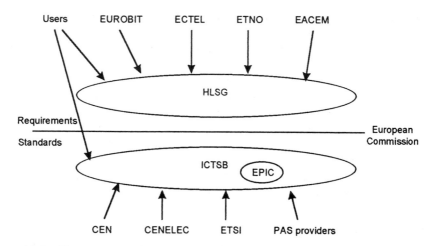

Figure 18.8   European standards structure [23]

(i) networks, e.g. access networks, network interfaces, internetworking, naming, addressing, numbering and routing, network management;

(ii) middleware, e.g. browsing and searching, systems management, security, fundamental processing services in end-user systems, middleware for human–computer interfaces, structured data files and metadata for multimedia information;

(iii) applications, e.g. medical informatics, libraries, electronic museums, road-transport infomatics, geographic information systems, ergonomics and industrial multimedia communications; and

(iv) architecture, e.g. framework of requirements and architecture, telecommunications services and network architecture, multimedia home platform and European culturally specific requirements.

18.8 Attributes of a top network/service provider in its attitude towards QoS

In this section we paint a picture of the attitude of a top network or service provider towards the quality of its services. To meet the challenges of the future, the successful network or service provider:

(a) is unlikely to provide a high quality service merely by specialising in the quality of its services. It will do so only if the quality of it services becomes part of its total-quality-management (TQM) programme. Management of quality of telecommunication services is a complex process and is unlikely to get maximum benefit unless all activities leading towards it are working at an optimum level;

(b) will listen to the customers and provide what they realistically expect at a price they are prepared to pay. Gone are the days when technocrats decided what was good enough for the customers. The top network or service provider of the future will work with the customers to identify their needs and develop the services to meet those needs;

(c) will publish achieved quality-of-service parameters which will be of concern to the customers. Failure to do so will be seen as a weakness and poor business practice. It will, at best, be seen as that of a network or service provider which is definitely not in the top league.

(d) will provide a high-quality service for its special interest group, i.e. the disabled, the poor and the less advantaged who depend on telecommunications for much of their relationships with the outside world;

(e) will respect the environment and contribute to its beauty by good ergonomical practices and careful use of resources, and will partake in the societal issues of the day of the community;

(f) will respect its customers and treat them as its primary source of income, and never forget that the customer may have a choice. A top network or service provider will, indeed, respect its customers, even if they have no choice;

(g) will keep abreast of innovation and constantly seek how new applications can be developed to enhance the quality of life of its customers;

(h) will take a leading role in the standardising process, if only to keep the industry informed and to assist it to work more efficiently;

(i) will work with, and not against, the regulators, and will advise the regulator how to improve its effectiveness in meeting the best interests of the customer;

(j) will provide quality information and advise to the Government on how to improve the quality of communications to the nation; and

(k) will recruit and train the best employees and treat them with respect and not as commodities, recognising that motivated employees produce the best productivity, quality of work and customer relations.

The above litany of qualities of a top network or service provider may appear idealistic. However, careful examination of the most successful companies will show that such companies exist and adopt many of the above criteria.

18.9 Conclusions

There is no doubt that the world is entering the information era. The fusion of communications, IT, the media and entertainment is spawning a plethora of new players each expanding into the others' competitive space. The global impact will be immense, changing for ever the traditional manner of communications. This chapter has outlined the drivers that are creating the information society, and has indicated the range of players that will be competing for a huge and expanding information market and how they are already positioning themselves in the value chain to reap maximum benefit. Emphasis has been given to this likely future environment because it will have a major impact on customer QoS requirements. These are bound to be more onerous than today.

Some views have been expressed on potential QoS drivers but it has not been possible to do other than speculate on the requirements. The information applications are still in the evolutionary stage and QoS requirements will evolve over time as new services and technologies are deployed.

18.10 References

1   SHWARTZ, P.: 'Post capitalist' (Wired On-Line Information, 1993)
2   STEWART, T. A.: 'The age of information in charts', *Fortune*, 4 April 1994
3   GORE, A.: Remarks at National Press Club, 21 December 1993
4   BANGEMANN, M.: 'Europe and the global information society'. Report of the High Level Group on the Information Society, May 1994
5   ETSI: 'Report on the sixth strategic review committee on European information infrastructure'. June 1995
6   EURESCOM: 'Shaping the information society – a view from European public network operators'. January 1997
7   OFTEL: 'Beyond the telephone, the television and the PC' (Office of Telecommunications, May 1995), consultative document
8   JAPANESE GOVERNMENT: 'Reforms towards the intellectual creative society of the 21st century'. Report, May 1994
9   PORTER, M. E.: 'Competitive advantage' (The Free Press, 1985)
10 VOGEL, H. L, 'Entertainment industry economics' (Cambridge University Press, 1994)
11 COOPER, M.: 'Human factors in telecommunications engineering', *Br. Telecom. Eng.*, 1994, **13**, (2)
12 WALKER, G., *et al.*: 'Interactive visualisation and virtual environments on the Internet', *Br. Telecom. Eng.*, 1996, **15**, (1)
13 WALKER, G.: 'Challenges in information visualisation', *Br. Telecom. Eng.*, 1995, **14**, (1)
14 LEGH-SMITH, J.: 'Navigating on-line service environments', *Br. Telecom. Eng.*, 1996, **15**, (1)
15 PRESTO, K.: 'From books to bytes — managing information in the information age', *Br. Telecom. Eng.*, 1996, **15**, (1)
16 SMITH, R.: 'Software agent technology', *Br. Telecom. Eng.*, 1996, **15**, (1)
17 ICC E100/1: ICC Project E100, draft summary record of the first plenary session, 14 September 1995
18 BOLERO USER ASSOCIATION LTD.: 1 Gainsford Street, London SE1 2NE, UK
19 BLUMFIELD and COHEN, 'An overview of the US telecommunications act of 1996' Technology Law Group. Available online from http://www.technology-law.com/techlaw/act_summary.html
20 DAVIC: Web site http://www.davic.org/
21 EUROPEAN COMMISSION (DGXIII), Round Table 3: 'Quality needs of electronic information and communications services', 1997 (see also Appendix 4)
22 MULTIMEDIA COMMUNICATIONS FORUM Inc.: 'Multimedia communications QoS', MMCF/95-010, June 1995
23 DICKERSON, K.: 'The needs of European information infrastructure standards'. IEE Colloquium on *European Collaborative Telecommunications Research – do we need it?*, 14 May 1997

Chapter 19

Standardising the framework

19.1 Introduction

The drawbacks of the lack of an internationally agreed architectural framework for the study and management of QoS have been discussed. Components of a framework have been described in Chapters 4–8. In this chapter these components are brought together and presented as a possible starting point for the standardisation body, such as the ITU-T, to be built into an internationally agreed architectural framework. Existing standards may then be mapped to the standard framework and refined, where necessary. Further work on new standards may also be based on the new standards framework.

A universally recognised framework, agreed in the international forum, could result in the following benefits:

(a) a clearer focus on the mapping of network technical performance to the end-to-end QoS of relevance to the customer and the service provider;

(b) consistency in the use of terms and definitions for describing quality criteria;

(c) comparability of QoS on networks and services throughout the world;

(d) an easier grasp of the significance and relevance of each parameter in the overall map of QoS;

(e) identification of all relevant QoS criteria and less likelihood of wasted or duplicated effort on similar topics, resulting in a clearer focus for individual studies leading to greater depth;

(f) the interests of various parties in the management of QoS, i.e. network providers, service providers and the users, could be analysed in the context of overall QoS, leading to an optimum number of performance specifications; and

(g) clearer identification of work areas could result in fewer resources and more output than the 'pre-architectural' era.

Potential users of the framework, illustrated in Figure 19.1, are the network and service providers, regulators and the regional standardisation bodies such as ETSI and ANSI, ITU-T, ISO and OECD.

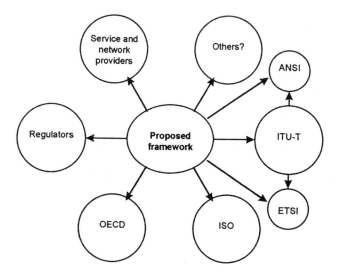

Figure 19.1   Potential users of the architectural framework

19.2 Architectural framework for the study and management of QoS

19.2.1 Criteria

An architectural framework should fulfil the following criteria:

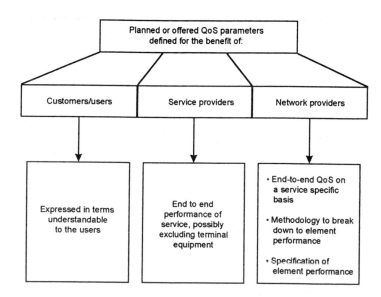

Figure 19.2   Parameter definition and specification of planned QoS

(a) It should be easy to understand and usable by all parties of tele-communications industry, in particular:

- users and customers;
- service providers;
- network providers.

Figure 19.2 shows how the interests of the three parties fit together.

(b) The framework must assist with the identification of most, if not all, of QoS criteria, network-related, non-network-related and the 'soft issues'.

(c) The framework must show the interrelationships of QoS between users of telecommunication services and the suppliers of such services (service providers and network providers).

(d) The framework must show the relationship between the QoS criteria and the service provider.

An architectural framework which fulfils the above requirements is illustrated in Figure 19.3.

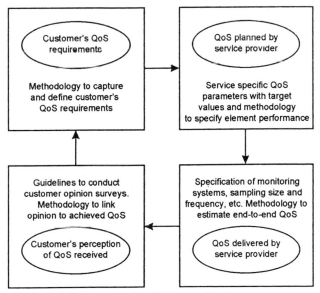

Figure 19.3   Architectural framework for the study of QoS

The principal feature of the framework is the logical division of the QoS-studies management into four viewpoints. The issues are therefore decoupled from those other issues which are not directly relevant. The salient points of each viewpoint and the relationship with the other viewpoint/s are dealt with in the following subsections.

19.2.2 Customers' QoS requirements

The study and management of customers' QoS requirements should contain the following functions:

- a matrix to facilitate the identification of QoS parameters in a consistent and standard manner;
- cell definitions of the matrix.

A matrix has been described in Chapter 4, with the cell descriptions in Appendix 2. Additional features to be associated with the matrix and the cell descriptions are:

(a) cell definitions must be generic;
(b) guidelines for the derivation of service-specific QoS criteria for the principal services used throughout the world should be considered;
(c) guidelines on the capture mechanism could be provided to interpret the effects of different cultures;
(d) principles of questionnaire design;
(e) guidelines for the selection of samples and sizes;
(f) guidelines to arrive at a meaningful set of criteria from the collected requirements.

Requirements expressed in the language of the customers may need to be translated, where necessary, into the language of the service provider for input to the decision-making process to determine the offered QoS.

19.2.3 QoS offered

(a) Parameters of offered (or planned) QoS for the benefit of users, service providers and network providers

Parameters of offered QoS need to be defined to express the basic QoS of principal services. These parameters should reflect end-to-end performance, and for the benefit of users, service providers and network providers. Any variations of definitions of performance for these three parties would depend on the service and the performance parameter being considered.

(b) Methodology for the breakdown to element performance

Guidelines for the breakdown of end-to-end QoS into individual network element technical performance would be needed.

(c) Specification of element performance

Standardisation of element technical performance is desirable.
     The standardisation of the two above categories would benefit equipment manufacturers. The planned QoS will form the basis for the specification of monitoring systems, in the next viewpoint.

19.2.4 QoS delivered

Two principal activities make up the QoS achieved or delivered by the service providers: first, the specification of monitoring systems, and secondly, the calculation of end-to-end QoS from the performance data of elements. The first activity is shown schematically in Figure 19.4 and described in Chapter 7.

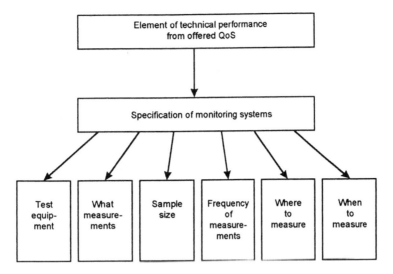

Figure 19.4 Specification of monitoring systems: constituent parts

Activities associated with the monitoring systems, all of which require study in the standardisation bodies, resulting in specifications or guidelines are:

- specification of test equipment for the different parameters;
- where the measurement is to be carried out;
- what measurements are to be carried out;
- sample size of measurements;
- where and how often (frequency) measurements are to be carried out.

Specifications of test equipment should not recommend any particular manufacturer, but state the technical principles and the characteristics of the measurements, e.g. the weighting curve of the psophometer.

The second activity, computation of the end-to-end performance based on measurements taken, is illustrated in Figure 19.5. The principal activities, each requiring study are:

(a) principles for the computation of end-to-end performance from element technical performance;
(b) algorithms for the computation of end-to-end QoS for more sophisticated parameters. An example is call quality to be computed from measurements taken on elements. These measurements may be a selection of noise, delay, slip, jitter and wander, all contributing to the transmission quality of the connection. The combined effect on a

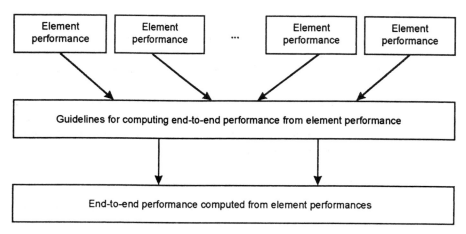

Figure 19.5   Computation of end-to-end performance from element performance

particular service is to be established and specified on a service-by-service basis; and

(c) where relevant, guidelines on the computation of confidence limits of the end-to-end performance arrived at by the combination of element performances.

19.2.5 Customers' perception of QoS

The principal study areas for customer perception of QoS are shown schematically in Figure 19.6.

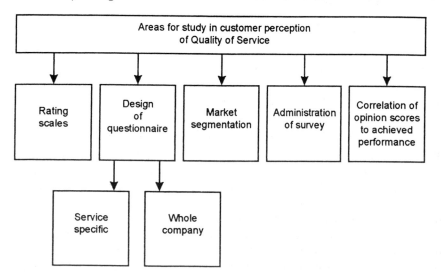

Figure 19.6   Study areas for customer perception of QoS

(a) Rating scales

Customer opinion is rated on scales, varying from 1 to 10. The following are some of the more popular scales used: 1–4, 1–5, 1–7, 1–10. Guidance may be given on qualitative descriptions of the intervals (see Figure 8.7) of the rating scale.

Studies may be carried out to choose which of these scales are most suited for both national usage and international comparisons. The standards forum could state the recommended ranges for the expression of customer perception ratings, stating the merits of differentiations, if any.

(b) Design of questionnaires

Generally there are two types of questionnaires: one to assess customer opinion of a particular service, the other to assess the overall opinion of a service provider. While it must be recognised that questionnaire designs for various business cultures could be different, generally applicable guidelines may be established and recommended by the standards body.

(c) Market segmentation

Market segmentation has, in general, followed industry classification. Individual experiences of service providers have also added further knowledge to the segmentation process. As competition in telecommunications is relatively new, regular reviews of existing knowledge could prove useful to both service providers and the regulators. It may not be wise for the standardisation body recommend how market segmentation ought to take place, and general guidelines may be sufficient.

(d) Administration of survey

As discussed in Chapter 4, surveys could be administered many ways. Studies to assess the methods of producing the most accurate customer survey are considered necessary in telecommunications. The standardisation body could provide guidelines for universal usage.

(e) Correlation of customer opinion rating to delivered performance

The correlation of customer opinions with delivered QoS forms the basis of the post mortem of any service provider to improve its performance. Before it carries out corrective action it is necessary to account for the discrepancy, usually present for many service providers, between the customer-opinion scores and the delivered performance.

Studies in this area could be initiated by both the service providers and user groups for a better understanding of customer behaviour. The standardisation body could provide certain guidelines, but the service provider should be left to carry out the detailed correlation.

19.3 Current standards and the new framework

With the development of a new architectural framework, all existing standards ought to be mapped to one of the four 'viewpoints' of QoS. Fine tuning of any of these standards should be considered to ensure that the function of the standard is maximised (see the example in Chapter 13). Further work may be carried out for new standards. Where drastic revisions or changes are necessary, the current standards may be completely rewritten to fit in with the new framework.

19.4 Conclusions

The following components, described elsewhere in this chapter, may be considered for standardisation:

(i)   the concept of the four viewpoints as illustrated in Figure 19.2;
(ii) the components of customer's QoS requirements, i.e. the matrix and the generic cell descriptions mentioned in section Chapter 4;
(iii) definitions of terms for expressing the offered (or planned) QoS of the principal telecommunications services (on a service-specific basis), as mentioned in Section 19.2.3;
(iv) principles of monitoring systems to include;

- nature of measurements, e.g. intrusive or nonintrusive for voice measurements;
- characteristics of measuring equipment;
- points in the network where measurements are to be carried out;
- frequency of measurements;
- sample size of measurements;
- time of measurements.

(v)   principles of calculation of achieved (or delivered) end-to-end QoS from measurements of individual element performance;
(vi) guidelines for carrying out customer-opinion ratings, e.g.

- selection of rating scales;
- selection of sample size;
- selection of samples;
- customer segmentation;
- administration of survey;
- publication of results.

(vii) principles of correlation of customer opinion ratings and QoS delivered; guidelines for interpreting and drawing conclusions.

The benefits of an architectural framework for the study and management of QoS should become more apparent when it is put to use. The benefits resulting from the investment of resources for the development of a framework and its subsequent adoption for further studies must be considered before further resources are expended on quality issues. Since more sophisticated service applications and issues regarding optimum economic levels of quality are likely to become more pertinent in the future, the time now is considered to be ripe for the development and adoption of an effective architectural framework.

Appendixes

Appendix 1: General management of quality

A1.1 Introduction

Total quality management (TQM) may be defined as philosophy and company practices which aim to harness the human and material resources in the most effective way to achieve the objectives of the organisation. TQM is basically geared for quality performance of every function in an organisation [1]. The characteristics of TQM are embraced in:

- universal participation;
- focus on customer needs;
- everything is a process which contributes to quality;
- continuous process improvement;
- better performance at lower cost; and
- excellence in communication and understanding.

Among the principal contributions made to this topic are the following 'gurus'.

According to Crosby [2,3]:

- quality is conformance to requirements;
- the system of quality is prevention;
- the performance standard is zero defects;
- the measure of quality is the price of nonconformance.

According to Deming [4]:

- create consistency of purpose;
- adopt the new philosophy;
- stop mass inspection;
- stop awarding contracts on price;
- find and solve problems;
- on-the-job training;
- redefine supervisor's role;
- drive out fear;

- breakdown departmental barriers;
- eliminate numerical goals;
- eliminate slogans and posters;
- have common terms and conditions;
- educate and retrain;
- top management co-ordination.

According to Juran [5]:

- build awareness;
- set goals for improvement;
- organise;
- provide training;
- carry out projects;
- report progress;
- give recognition;
- communicate results;
- keep score;
- continuous improvement.

The total quality organisation:

focuses on:

- continuous process improvement;
- everything as a process;
- the use of scientific methods;
- perfection as the goal;

through

- universal participation;
- everyone;
- everywhere;
- individuals and teams;

resulting in

- customer satisfaction;
- exceeding expectations;

for

- internal customers; and
- external customers.

More than anything else, what distinguishes a TQ organisation from an ordinary one is the way its people think and act. The value that people place on quality of performance in very activity and what they do to improve the quality of their work are key factors in a TQ organisation. It is vital that the organisation culture supports the total-quality concepts.

For further information on TQM and its implementation readers are recommended to refer to published literature [2,3,4,5,6].

A1.2 Contributions from standardisation bodies

The relevant standards pertaining to management of quality published by the ISO are:

- ISO 9000   [7]
- ISO 9001   [8]
- ISO 10011   [9]

The first two deal with the development of processes and their documentation. These are very useful in the documenting of processes within the telecommunication-service provider's organisation for the management of quality. The ISO 10011 standard provides guidelines for the audit of the processes. These ISO standards, when followed, will help to ensure that the processes are logical and written in a proper manner. They do not ensure that the processes are good or optimum for the purpose. The European Foundation for Quality Management (EFQM) model and the ISO standards on quality management and audit adequately cover the fundamental principles for the management of a service provider's organisation. What is also required is a focused micromanagement of quality of service within the macroenvironment of TQM.

National standardisation bodies such as the British Standards Institution (BSI) also have standards on TQM [10].

A1.3 Contribution made by EFQM

Recognition of the importance of TQM is acknowledged by the formation of EFQM [11] in 1989. It was founded by 14 large European companies to promote the practice of TQM and to cross-fertilise the quality management ideas applied across many industry sectors. There are more than 300 companies which are members. Regular meetings are held to promote learning and the practice of quality management. It has the backing of the European Commission. European telecommunications-service providers which are members include BT, France Telecom, Deutsche Bundespost Telekom, Telecom Italia and Telefonica.

Figure A1.1 illustrates a model, showing the breakdown of effort put into the organisation and the resulting output of that organisation. If TQM is practised as it should be, and if the organisation is working at peak efficiency, then the enablers and the results, illustrated in Figure A1.1, should each be awarded the maximum of 500 points. The number of points awarded is indicative of how efficiently the organisation has applied TQM principles. Guidelines for the allocation are given by the EFQM.

In the EFQM model, if the 'enablers' identified in the diagram are carried out according to the principles of TQM then the 'results', as identified in the results half of the model, should produce maximum benefit. These benefits are measured by the points given to the various categories. Thus customer satisfaction, if maximised, should be awarded 200 out of 500 points. This is

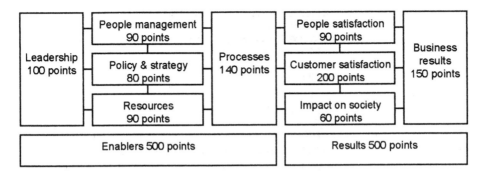

Figure A1.1   The EFQM model for rating performance of an organisation

analogous to quality of service. Impact on society, worth 60 points, could also be included in the quality of service. It is clear that at least 40% of the results of any organisation are measured by the resulting effects of quality of service on the customers. This model and the work of the organisation continue to be a source of inspiration not only to telecommunication companies but also to companies in other industry sectors.

The principal contribution made by the EFQM, apart from the model and the award to successful companies practising TQM, is the encouragement given to its member companies to apply the successful practices across industry sectors. EFQM is a respected organisation within Europe and its contribution could be noted by all service providers for their own benefit. Detailed examination of its work is outside the scope of this book and for more detailed information on its activities and achievements the reader is advised to contact its office in Belgium [11].

A1.4 References

1   HAIGH, P.: 'Total quality management'. Quality Plus, Monks Risborough, Bucks, UK
2   Crosby, P. B.: 'Quality is still free: making quality certain in uncertain times' (McGraw Hill, 1996)
3   CROSBY, P. B.: 'Quality is free: the art of making quality certain' (McGraw Hill, 1979)
4   DEMING, W. E.: 'Out of the crisis; quality productivtity and competitive position' (Cambridge University Press, 1986)
5   JURAN, J. M.: 'Juran on quality by design: the new steps for planning quality into goods and services' (Free Press, 1992)
6   OAKLAND, J.: 'Total quality management: the route to improving performance (Butterworth Heinemann, 1993)
7   ISO 9000: 'Quality management and quality assurance standards' (International Standards Organisation, 1994)
8   ISO 9001: 'Quality systems. Model for quality assurance in design, development, production, installation and servicing' (International Standards Organisation, 1994)
9   ISO 10011: 'Guidelines for auditing quality systems. Part 1: Auditing; Part 2: Qualification criteria for quality system auditors.; Part 3: Management of audit programs' (International Standards Organisation, 1993)

10 BS 7850: 'Total quality management. Part 1: Guide to management principles; Part 2: Guidelines for quality improvement' (British Standards Institution, 1992, 1994)
11 European Foundation for Quality Management, Avenue des Pleiades 19, B-1200 Brussels, Belgium

Appendix 2: Matrix to facilitate capture of QoS criteria: cell descriptions

A matrix was described in Chapter 4 to facilitate identification of QoS criteria for telecommunication services. In this appendix the cell descriptions are given; these are generic QoS criteria. Service-specific QoS criteria are derived from these, wherever applicable. The matrix is reproduced here for convenience.

A2.1 Sales and precontract activities

A2.1.1 Speed

Speed of sales may be indicated by the time taken from the initial contact between the customer and the service provider to the instant an effective contract is placed for a service. Not all contacts will result in sales. Where no sales results, the speed of sales will be the time elapsed from initial contact to the instant an offer is made by the service provider after all the pertinent information has been supplied to the customer.

A2.1.2 Accuracy

Accuracy is exemplified by the correctness and completeness in the description and delivery of all relevant service information, normally expected by the customer before effective contract, e.g. service features, performance, charges, service support, provision, time etc.

A2.1.3 Availability

This criterion deals with access to the service information from the service provider. The mode of access may be in writing, electronic, verbal, or in person. Also included could be the number of offices, hours these are open and hours the staff may be accessed for information.

A2.1.4 Reliability

Reliability of sales (information) is an indicator of how well the service provider has satisfied the customer in terms of precontract formalities. The degree of accuracy associated with the information, together with how easily this information is available over a given period, would determine how this parameter has performed. It is also an indication of the professionalism of the members of the service provider serving the customers.

A2.1.5 Security

Confidentiality requirements of customers from the service provider on all activities related to sales form this quality criterion. This quality criterion in association with precontract activities, is likely to be of concern during special

Service function \ Service quality criteria	Speed 1	Accuracy 2	Availability 3	Reliability 4	Security 5	Simplicity 6	Flexibility 7
Service management sales & precon-tract activities 1							
provision 2							
alteration 3							
service support 4							
repair 5							
cessation 6							
Connection quality connection establishment 7							
information transfer 8							
connection release 9							
Charging & billing 10							
Network/service management by customer 11							

Figure A1.2   Matrix to facilitate determination of QoS criteria

situations, e.g. dealings with government agencies and where commercial confidences are of importance.

A2.1.6 Simplicity

This concerns the ease with which all activities associated with sales (and purchase) may be carried out with the service provider. Included are: ease of identification of the point of contact for sales, ease with which information supplied is understandable, the ease with which forms can be filled and ease with which orders can be placed.

A2.1.7 Flexibility

The flexibility with which the service provider will accommodate the individual customer's requirements is identified in this performance measure.

A2.2 Provision of service

A2.2.1 Speed

Speed of provision is indicated by the time taken from the instant of effective contract to the instant service is available for use by the customer.

A2.2.2 Accuracy

The correctness and completeness in the provision of any service and associated features specified or implied in the contract.

A2.2.3 Availability

This depends on access to resources by the service provider to meet provision of service or product agreed with the customer in the contract. It could also deal with the facility offered by the customer for the installation of the service.

A2.2.4 Reliability

This concerns the long-term performance of the above three performance criteria, e.g. over a period of one year.

A2.2.5 Security

This concerns confidentiality requirements with regard to a service. Examples include:

- ex-directory-number-protection facilities;
- special requirements for government installations;
- compliance with Data Protection Act;
- nondisplay of calling-line identity etc.

A2.2.6 Simplicity

The ease and convenience with which a service can normally be expected by a customer after effective contract, e.g. minimum inconvenience to customer during installation, ergonomically designed cabling and fittings, simple but effective installation practices etc.

A2.2.7 Flexibility

This concerns options normally expected by the customer to accommodate special requirements on the provision of the service without departing from the terms of the contract. Examples include timing of the provision of the service to suit the customer, provision of terminal equipment to match customer preferences (e.g. appointments) where possible etc.

A2.3 Alteration to a service

A2.3.1 Speed

This is the time taken from request to a service provider for an alteration to a service to the instant alterations are incorporated and is available for use.

A2.3.2 Accuracy

This covers the correctness and completeness with which requests for alteration to service are carried out as specified or implied in the contract.

A2.3.3 Availability

This concerns access to resources by the service provider to carry out alteration to the service as requested by the customer and implied in the contract.

A2.3.4 Reliability

This is indicated by the long-term performance of the above three measures, e.g. over a period of one year.

A2.3.5 Security

This concerns customers' confidentiality requirements with regard to all aspects of alterations carried out on a service. For example, a customer may require ex-directory listing which may involve a change in number. It may involve a change in status of the working of the organisation, with possibly the award of a secret contract from the government requiring secure communications.

A2.3.6 Simplicity

This concerns the ease and convenience with which alteration(s) requested may reasonably be expected by the customer.

A2.3.7 Flexibility

This concerns customer requirements on the options to accommodate special requirements relating to alteration of a service. Examples include accommodating the customer's request for meter reading at a requested time when moving to a new address, and capability to accommodate a customer's request to carry a telephone number to a new address.

A2.4 Service support

A2.4.1 Speed

This criterion may be expressed as the time elapsed from the instant a customer requests service support to the instant support is provided to the satisfaction of the customer.

A2.4.2 Accuracy

This concerns the correctness and completeness in the service support as specified or implied in the contract.

A2.4.3 Availability

This concerns the presence of access facilities for the customer making service-support requests for the customer. For the service provider, it is the resources to provide the service support.

A2.4.4 Reliability

This is determined by the long-term performance of the three previous criteria, e.g. over a period of one year.

A2.4.5 Security

This covers customers' confidentiality requirements from the service provider on all matters relating to requests and provision of service support, e.g. security considerations on service support for special situations such as government installations. There may be a need for any service support to be kept away from the public eye. The requirement would be specified by the customer.

A2.4.6 Simplicity

This concerns the ease and convenience with which customers can normally expect to request and receive service support and any associated activities.

A2.4.7 Flexibility

This concerns customers' requirements on ease and convenience with which service support may be requested and provided. Examples are varying levels of service support and various modes of access. Service support may be obtained verbally, in print or by electronic mail. In other cases a flexible approach may be provided, e.g. credit card payments of bills over telephone.

A2.5 Repair

A2.5.1 Speed

This criterion may be expressed by the time taken from the instant repair is requested to the time it is carried out and all service features are restored to normal use.

A2.5.2 Accuracy

This concerns the correctness and completeness of the repair carried out as agreed or implied in the contract.

A2.5.3 Availability

This concerns the presence of facilities for making requests for the repair service. These facilities would include hours of access as well as methods of access. For the service provider, it would mean presence and access of resources to carry out the repair.

A2.5.4 Reliability

This is determined by long-term performance of the three preceding performance criteria, e.g. over a period of one year.

A2.5.5 Security

This concerns a customer's requirements on confidentiality in matters relating to repair service. For example, this would include sending of approved staff for defence establishments and respecting a customer's information on installations not being divulged to third parties.

A2.5.6 Simplicity

This concerns how simple and effective is the manner in which repairs are carried out. For example, if one person is adequate to carry out repairs it would be most unprofessional if three people turned up and caused the customer annoyance.

A2.5.7 Flexibility

This concerns customers' requirements on the options available in carrying out repairs. For example, repairs may be carried out, where possible, in the first instance, without access to customer premises or where possible remotely. Repairs may also be carried out at the customer's convenience should entry to premises be required. Alternative service may be offered if service is unusable.

A2.6 Cessation

A2.6.1 Speed

This criterion may expressed by the time taken from request for cessation of service to the instant it is carried out.

A2.6.2 Accuracy

This concerns the correctness and completeness in carrying out the cessation of a service and the associated activities, depending on whether the cessation was initiated by the customer or the service provider.

A2.6.3 Availability

This concerns facilities offered to customers for making requests for cessation of service. For the service provider, it concerns access to resources to carry out cessation of a service. This could include issuing of final bills and closing of accounts and dealing with the correspondence for the particular service.

A2.6.4 Reliability

This depends on the long-term performance of the three preceding performance criteria, e.g. over a period of one year.

A2.6.5 Security

This concerns customers' confidentiality requirements regarding all activities connected with the cessation of service. For example, a business may state that cessation of services to its premises should not commence until a move to new premises is completed and the new communications services are working satisfactorily.

A2.6.6 Simplicity

This concerns customers' requirements on the ease and convenience with which activities associated with the cessation of a service are carried out. For example, a customer may wish to close an account with the service provider in one transaction.

A2.6.7 Flexibility

This concerns customer requirements to minimise inconvenience during the process of cessation of a service, for example, wrong cessation due to a error of the service provider, cessation requested by the customer to be carried out at a time requested by the customer, e.g. a specified hour to coincide with moving office or residence.

A2.7 Connection establishment

A2.7.1 Speed

This is given by the time elapsed from the input of the last address digit to the instant the signal is received from the network to indicate the status of the called party.

A2.7.2 Accuracy

This measure deals with:

(i)   the correct and complete indication of the status of the called party when it has been reached by the address digits; and
(ii) reaching the called party identified by the address digits.

A2.7.3 Availability

This is the probability of a customer being able to establish a connection when requested. This parameter could be expressed additionally with (i) frequency of unavailability and (ii) maximum duration of any one unavailable period.

A2.7.4 Reliability

This is determined by the long-term performance of the above three parameters over a period of one year.

A2.7.5 Security

This concerns a customer's confidentiality requirements on connection attempts, for example the display of calling-line identity to the called party against the stated wishes of the customer.

A2.7.6 Simplicity

This concerns the customer's requirements on the ease and convenience with which connections may be established. Examples are:

(i)   personalised numbering schemes;
(ii) short dialling codes;

(iii) easily identifiable blocks of numbers for uniquely identifiable blocks of customers;

(iv) providing easy-to-understand network reactions, e.g. tones and announcements; and

(v) user-friendly protocols during connection set-up.

A2.7.7 Flexibility

This concerns customers' requirements in the options available in the process of setting up connections. Examples include various network options when connection to a called party is not possible, e.g.

(i) call-forwarding facility offered to customers whereby the called party can transfer the call to another attended number in the event of the main number being unattended;

(ii) ring-back-when-free and/or call-waiting indication options when the called party is engaged; or

(iii) voice message, giving reason for call not maturing.

A2.8 Information transfer

A2.8.1 Speed

This is measured by the rate at which information is transferred from calling to called end (where relevant) or the time taken to transfer the information from the transmitting to the received end.

A2.8.2 Accuracy

This is the degree of faithfulness of the received information to the information sent over a connection. For example, for speech over the telephony connections, this is expressed by 'call clarity' or 'call quality'.

A2.8.3 Availability

This quality criterion is expressed by the probability that the established call will hold for the intended duration of the connection. This parameter is, strictly, to be separate from the availability at connection set-up.

A2.8.4 Reliability

This is determined by the long-term performance of the above three criteria, e.g. over a period of one year.

A2.8.5 Security

This concerns a customer's confidentiality requirements during the information-transfer phase. For example, a customer could be concerned about the security against other parties 'listening in' over a mobile phone.

A2.8.6 Simplicity

This measure includes the ease and the convenience with which interactions with the networks may be undertaken by the customer, when interactive service is provided. Examples include:

(i) ease with which additional codes may be inserted by the customer during the information transfer for optional services and facilities;

(ii) over-riding of announcements by input digits for the benefit of customers who are familiar with the use of service; and

(iii) user instructions may be made simple where interactive services are offered.

A2.8.7 Flexibility

This concerns customers' requirements on the options available for the use of the service. Examples are:

(i) call-waiting indication could offer visual or audible (or both) indications to the user (with facility to select a combination); and

(ii) options available on the customer-network interactions.

A2.9 Connection release

A2.9.1 Speed

Customers may have specific requirements for the connection-release times to make them compatible with the terminal equipment they intend to use with the network.

A2.9.2 Accuracy

Customers may have requirements on faithfulness to specified logic-of-release protocol and accuracy of network-release times of a connection.

A2.9.3 Availability

The release should always take place when requested. The customer would be concerned that charges may be incurred if calls are not cleared as requested.

A2.9.4 Reliability

The long-term performance of the above three parameters, e.g. over a period of one year.

A2.9.5 *Security*

Customers' confidentiality requirements. These would comprise information on the call or clearing of the call made available to a third party during clearing procedure.

A2.9.6 *Simplicity*

Understandable and easy procedures for the release of an established connection.

A2.9.7 *Flexibility*

Capability of the service provider to offer options on the connection-release times. For certain applications this could be important.

A2.10 *Charging and billing*

A2.10.1 *Speed*

There are two measures, one for charging and other for billing.

Speed of charging may be expressed by the time taken for the charging information to be made available to the customer after the charge has been incurred. Examples of instances where this might be required are:

- advice of charge after a call is made;
- advice of current cumulative charges;
- advice of charges up to a certain period.

Speed of billing is expressed by the time taken for a bill to be presented to the customer from the date of request for a bill by the customer.

A2.10.2 *Accuracy*

The completeness and the accuracy of the billing information in reflecting actual use of the service. The tolerance or the limits of this accuracy may be specified by the maximum number of errors (expressed as a percentage) *and* the magnitude of the largest error.

A2.10.3 *Availability*

It is expresed by the probability of specific billing information being available to the customer. To the service provider this quality criterion could also mean access to resources to establish charging information and produce bills.

A2.10.4 *Reliability*

Long-term performance of the three above criteria, e.g. over a period of one year.

A2.10.5 Security

Customers' and service providers' requirements on the security of charging, e.g. security of the network against fraud. An example of fraud is the ability to make unpaid calls on pay phones. Customers would wish to ensure that no other customer(s) can incur charges against their or a specified account number.

A2.10.6 Simplicity

Customer's requirements on easy-to-understand billing information and formats.

A2.10.7 Flexibility

Customers' requirements for options available on the:

(i)   format of the bills; and
(ii) frequency and date when billing information may be available.

A2.11 Network/service management by customer

A2.11.1 Speed

This quality criterion deals with access and response times when exercising the network/service-management function by the customer.

A2.11.2 Accuracy

The correctness and completeness of execution of a network- or service-management request, as specified or implied in the contract, e.g. accuracy of management information on call records.

A2.11.4 Availability

This quality criterion may be expressed by the probability of availability of the network/service management facility when required for use by the customer. It may also be specified by the outages as specified for the 'connection establishment – availability' cell specified in A2.7.3.

A2.11.4 Reliability

The long-term performance of the three preceding criteria, e.g. over a period of one year.

A2.11.5 Security

All matters related to confidentiality requirements on the correct management of network-service-management facilities and functions (e.g. use only by authorised users/customers).

A2.11.6 Simplicity

This criterion deals with user friendliness of the network or service-management facilities offered to the customers.

A2.11.7 Flexibility

This criterion deals with customers' requirements on the options for customisations in the network(s)/service-management facilities.

Appendix 3: Sample questionnaire to capture customers' QoS requirements

In the practical example of managing QoS for basic telephony given in Chapter 17, a questionnaire was mentioned. In this Appendix a sample of such a questionnaire is given.

Part 1

1 Precontract activities

(i)   What is the maximum time in which you would expect answers to all your enquiries on matters related to the supply of the basic telephony service and other precontract activities to be completed with the service provider?

_____ (days) _____ (hours) _____ (minutes)

(ii) On a 1–10 scale, please indicate the *relative priority* for this parameter in relation to the other parameters.

Lowest									Highest
1	2	3	4	5	6	7	8	9	10

Tick to show priority

2 Provision

(i)   What is the maximum period in which you would expect the service provider to install the basic telephony service from instant of effective contract?

_____ (days) _____ (hours) _____ (minutes)

(ii) On a 1 – 10 scale, please indicate the *relative priority* for this parameter in relation to the other parameters.

Lowest									Highest
1	2	3	4	5	6	7	8	9	10

Tick to show priority

3 Time to repair

(i)   What is the maximum time (from the instant a fault is reported) in which you would expect repairs to be carried out?

_____ (days) _____ (hours) _____(minutes)

(ii) On a 1–10 scale, please indicate the *relative priority* for this parameter (in relation to the other parameters).

Lowest									Highest
1	2	3	4	5	6	7	8	9	10

Tick to show priority

4 Time for resolution of complaints

(i) Please state the maximum time for the resolution of complaints.

_____ (days) _____(hours) _____(minutes)

(ii) On a 1–10 scale, please indicate the *relative priority* for this parameter (in relation to the other parameters).

Lowest									Highest
1	2	3	4	5	6	7	8	9	10

Tick to show priority

5 Time for establishing a connection

(i) While making a call, what is the maximum connection set-up-time you would be prepared to tolerate for a national call and for an international call?

National call _____ (seconds)

International call _____ (seconds)

(ii) On a 1 – 10 scale, please indicate the *relative priority* for this parameter (in relation to the other parameters).

Lowest									Highest
1	2	3	4	5	6	7	8	9	10

Tick to show priority

6 Misrouted calls

The service provider will attempt to provide 100% correct routing of calls. However this may not always be possible. What is the maximum number of misrouted calls you will be prepared to tolerate in one year?

(i)   Maximum number of misrouted calls
      _____ (please quote a figure)

(ii) On a 1–10 scale, please indicate the *relative priority* for this parameter (in relation to the other parameters).

Lowest									Highest
1	2	3	4	5	6	7	8	9	10

Tick to show priority

7 Call quality

The service provider will attempt to provide 100% of connections of sufficient quality to enable you and the person at the other end of the connection to understand each other without difficulty. However, this may not always be possible. How many calls, in one year would you be prepared to tolerate having moderate difficulty in understanding the other person?

(i)   Maximum number of calls with moderate difficulty in understanding the other person
      _____ (please quantify)

(ii) On a 1 – 10 scale, please indicate the *relative priority* for this parameter (in relation to the other parameters).

Lowest									Highest
1	2	3	4	5	6	7	8	9	10

Tick to show priority

8 Availability of service to make a call

It would be prohibitively expensive to provide 100% availability of network resources for 100% of the time.

(i)   Please state the minimum outage performance you require.
      Maximum number of outages in one year
      _____ (please quantify)
      Maximum duration of any one outage
      _____hours_____minutes_____seconds (please quantity)

(ii) On a 1 – 10 scale, please indicate the *relative priority* for this parameter (in relation to the other parameters).

Lowest									Highest
1	2	3	4	5	6	7	8	9	10

Tick to show priority

9 Continued availability for the intended duration of the call after the connection has been established

(i) The network resources may sometimes malfunction and an established connection may disconnect before the intended call release. Please state the maximum number of unintended connection releases you would be prepared to tolerate in one year.
Maximum number of connection releases your are prepared to tolerate
_____ (please quantify)

(ii) On a 1–10 scale, please indicate the *relative priority* for this parameter (in relation to the other parameters).

Lowest									Highest
1	2	3	4	5	6	7	8	9	10

Tick to show priority

10 Accuracy of bills

(i) The service provider will attempt to ensure that your bills are 100% accurate. However, there is an extremely small risk of mistakes. Please state what the maximum error you would be willing to tolerate if undetected by yourself or by the service provider.

Maximum number of errors in one year
_____ (please quantify)

Maximum magnitude of any one error
_____ (please quantify in units of currency)

(ii) On a 1–10 scale, please indicate the *relative priority* for this parameter (in relation to the other parameters).

Lowest									Highest
1	2	3	4	5	6	7	8	9	10

Tick to show priority

Part 2

Other quality criteria not captured by the questionnaire
Please state here any quality criteria you consider important, but are not covered in the questionnaire, and indicate the levels of performance you would expect.

_____

_____

_____

Part 3

Report on quality achieved by service provider for telephony service
(i) Please state the quality parameters which you wish the service provider to include in published report on achieved performance for the telephony service. Please state parameters from Part 1 of this questionnaire and/or state your own parameters. Please state as many parameters as you consider necessary.

Parameter 1_____

Parameter 2_____

Parameter 3_____

Parameter 4_____

Parameter 5_____

Parameter 6_____

Parameter 7_____

Parameter 8_____

Parameter 9_____

Parameter 10_____

(ii) Please state desired frequency of the report and the reporting period (e.g. quarterly for periods Jan-Mar, Apr-Jun, Jul-Sep, Oct-Dec).

Frequency _____

Period _____

Appendix 4: Illustration of service-specific issues: customers' requirements on electronic-mail service

To illustrate the range of quality issues to be addressed, a list identified by a recent study by the European Commission is given here[†]. Study of the requirements shows that users can be very demanding on quality matters. The list also shows that a wide range of issues are to be addressed for lasting satisfaction among users of electronic mail (e-mail), which is a relatively simple information service. The list is intended as a guide to future service providers, network providers and regulators, for identification of future QoS issues.

The list of quality requirements for electronic mail includes:

A4.1 Service provision

1   The service provider shall specify the service features, whenever possible, using the terms and definitions in the relevant ITU-T Recommendations.

   The service provider is free to specify the service features in simple and plain language to the customers. However, if customers ask for formal specifications, offerings must be stated in recognised terms used in relevant ITU-T Recommendations. The availability or otherwise of the following service features shall be supplied on request by the user/customer:

   (i)   transaction time
      • nominal
      • what was achieved in the recent past;
   (ii) indication of whether or not mail has been delivered;
   (iii) express mail – whether a mail could be delivered by a specified time by the sender;
   (iv) receipt that the mail has been read;
   (v)   probability of misrouteing;
   (vi) when the service provider analyses the e-mail for volume, performance etc. it should make public what analysis have been carried out. Customers should be made aware of the *nature* of the analysis to satisfy their concerns on traceability of their mail. The service provider need not make known the *findings* of the analysis.

2   Information sought by potential and existing customers on the e-mail service feature (e.g. tariff, conditions of use etc.) shall be provided by the service provider in any of the following forms:

      • in writing
      • electronically
      • by fax

[†] Interim report on 'Quality needs in electronic information and communications services', presented to ICT Partnership meeting DG XIII of European Commission, at Brussels, in July 1997.

- by telephone
- in person

The supplier shall provide the answers to customers' queries in any of the above forms if requested, unless this is impossible, (e.g. in person when the service provider is not open to the public).

3   The service shall be provided by the service provider at a time to be specified by the customer. It should be possible for the customer to sign on electronically for the service stating the time at which the service becomes effective.

4   Customers should be offered the facility to cease an e-mail service at any time including a specified time in the future.

A4.2 Despatch and receipt of mail

5   The time for response when customers are accessing their mail, and any other interactions, should be as low as possible. The service provider should specify typical speeds and what portion is attributable to the telecommunications company and what portion is attributable to the computers and systems of the service provider.

6   Customers shall have the facility to compose and edit the contents of the mail off-line and send it whenever they wish.

7   Where the facility is offered for a recipient of mail to acknowledge receipt, having read mail as opposed to storing without reading, these types of acknowledgements shall be at the discretion of the recipient. Where recipients have made it clear that they do not wish to acknowledge a message, this fact must be made available to the senders.

8   Customers shall have the facility to store mail in categories to be specified by them. Storage shall be free for a certain period of time. The maximum size that could be stored must also be specified.

9   If unsolicited mail causes annoyance to the recipient, the recipient should be able to bring the matter to the attention of the service provider. The service provider should be able to warn the sender of the matter and, if unjustified despatch of unsuitable material continues, the service provider should have the capability to remove the offender from the system.

10   The recipient must be advised whether the mail received is complete and without error.

11   Where the recipient has requested anonymity, and if the sender chooses to send a copy of mail to this recipient, then the anonymous recipient's details shall not appear in other addressees' mail, on either the original or other copies.

12   Recipients shall be offered the facility to interrogate how many mails are waiting for them and, at a second level of request, determine who these are from and, at a third level, display the full mail.

13   The customer to be provided with a set of simple statistics of mails received during a certain period, e g the billable period.

14 Service providers should develop and provide, with the collaboration of telecommunication-network providers, a paging message (bleep) when a mail has reached the recipient, thus prompting the customer to read. This facility may be restricted or categorised into normal and express mails.

15 E-mail should be made suitable for users to send legally binding contractual documents. This means that the delivered version of the text should be an exact copy of that sent. Additionally, only the recipient must receive a copy of such a document. The role of encryption should be for secrecy between the sender and the receiver and not be necessary for correct transmission.

16 A facility would be useful for a pager to be activated for various classes of mail received, e.g. express or ordinary mail, upon a mail being delivered. This would circumvent the user having to search for mail. When mail is delivered a pager indication may show that mail is waiting to be read. This must be optional and be under the control of the recipient.

17 The user should have the following mail options:
- receive only
- send and receive option but the receive option, when the recipient is absent, should be delivered by the provider by telephone, fax or post to an address nominated by the recipient.

18 The recipient should be able to download mail from a public library where access may be made available from a public computer.

19 The user should have the facility to mask all the routing information from a mail.

20 E-mail should have negative or positive filtering, i.e. the facility by the user to specify the addresses from which mail may be received or barred.

21 All communications (text, graphs, tables etc.) to be received without distortion.

A4.3 Ease of use

22 The address format should be easy to;
- remember
- search for in the e-mail directory.

23 The software used for the e-mail shall have user-friendly protocols for access, usage and release phases. The protocols shall be unambiguous and take into consideration, where possible, the local culture. Additionally, they shall also take into consideration the type of terminal equipment used by the customers.

24 There should be interoperability;
- between service provider and service provider, and
- between service provider and customer.

25 The sender's name and address is to be included in the mail.

26 Reason for nondelivery or late delivery of message must be indicated.

27   Advise, when requested, that the whole file has been delivered to the distant end.

A4.4 Service support

28   Technical and nontechnical after-sales support for the e-mail service shall be made available during periods of the day and week to reflect the local business and cultural requirements. Service provider could provide a choice of access for customers to seek help, e.g. by telephone, in person, by fax, by post and by electronic means. Technical assistance to nontechnical customers shall be provided in as simple and user-friendly language as possible. Consideration ought also to be given to customers with special needs, e.g. the partially deaf or wholly deaf, disabled etc.

29   The speed of response and resolution of queries by customers shall reflect the local custom.

30   The technical help line must be effective.

31   Instruction manuals for use of the software and all associated guidelines in the operation of the e-mail must be user friendly.

32   When there has been a service outage attributable to the service provider (and not the telephone company), the reason for the outage and the implications, if any, to customers who may have sent mail, shall be supplied, free-of-charge, to the customers at the resumption of normal service.

A4.5 Security

33   There shall be a facility, upon request, of confidentiality of e-mail address on similar lines to ex-directory telephone numbers. The service provider should ensure that the incoming mail is delivered only from those whose addresses have been given by the customer.

34   The software for e-mail should not limit the applications of any existing software on the computer, whether or not e-mail related.

35   To ensure delivery of mail, the service provider's computer should store information until acknowledgement from the far end has been received that the mail has been received and read.

36   The service provider shall ensure that the customer's personal details will not be divulged either intentionally or unintentionally to a third party.

37   The service provider must take steps to ensure that any financial information on the customer, e.g. credit-card details, shall not be made available intentionally or unintentionally to a third party.

38   The service provider must take steps to ensure that only the addressee can read the mail and that no one else can have access to its mail.

39   E-mail software should offer the facility for customers to access their mail from any other copy of the same software on any other computer anywhere in the world but through a system of passwords.

40   If, during the downloading of mail by a customer, an outage occurs, the mail should be preserved by the service provider until it has been read and acknowledged by the customer.

41 A facility for virus check should be made available. Any message delivered to the user should have been checked for virus by the provider. This may raise a legal issue on privacy.

A4.6 Compatibility

42 The far end should advise the customer whether the message can be downloaded.

43 When a customer sends a mail, the system shall not exhibit 'loop automatic reply' resulting in a number of to-and-fro messages.

44 The conditions under which customers can enter into dialogue e-mail conversation, if available, should be specified. If such a facility is available, the time to reach the called end should be specified.

45 Electronic-mail programs should feature a standardised interface to allow other software to interwork easily with them (e.g. en/decoders, signatures etc.)

A4.7 Tariff, charging and billing

46 All categories of charges to be clearly indicated.

47 Where there are many tariff options, customers should be given the option of changing to another after a reasonable period of time. Any qualifying time is to be specified in the terms and conditions of the service.

48 One option for payment should be zero payment to the service provider for standard service. The service provider should collect the fee from the telecommunications company which provides the access to the service provider.

49 Facility to be made available for the customer to access an accrued charge from the previous chargeable period.

50 If free access time is allowed, the customer should be offered the facility to ascertain the amount of time used, or the amount of time still available free, for the current period.

51 The service provider shall offer the customers the choice of payment by nonelectronic means, i.e. by standing orders through the bank, cheques by post on a regular basis etc. There should also be choice for the customer to change the method of payment, provided that requests for change are not too frequent. The choice of payments methods should be stated in the terms and conditions of service.

Index